Laser Technology in Chemistry

One-Day Symposium in Stockholm
on November 10, 1987
arranged by the Royal Swedish Academy
of Engineering Sciences (IVA)
in coordination with the Swedish Board
for Technical Development (STU)

Edited by Hans Medin and Sune Swanberg

Springer-Verlag Berlin Heidelberg GmbH

Hans Medin
Royal Swedish Academy of Engineering Sciences
P. O. Box 5073, S-102 42 Stockholm, Sweden

Sune Swanberg
Department of Physics, Lund Institute of Technology
P. O. Box 118, S-221 00 Lund, Sweden

This book originally appeared as an issue of the journal
Applied Physics B: Photophysics and Laser Chemistry
Volume 46 Number 3
(ISSN 0721-7269) © Springer-Verlag Berlin Heidelberg 1988

ISBN 978-3-540-50132-9   ISBN 978-3-642-52476-9 (eBook)
DOI 10.1007/978-3-642-52476-9

Laser Technology in Chemistry

Editorial

Laser Technology in Chemistry

The remarkable development of laser technology and the rapidly increasing knowledge of the interaction between light and atoms/molecules are having a major impact on chemistry. Laser spectroscopic techniques provide new means for a detailed understanding of chemical reactions. Analytical spectroscopy based on lasers is characterized by an extreme sensitivity and selectivity in chemical analysis. New diagnostic techniques enable powerful, nonintrusive measurements in aggressive chemical environments, for example, in connection with combustion and semiconductor processing. Laser-induced chemical processes provide new and powerful means of producing certain chemicals as well as new semiconductor processing technology. The extreme wavelength selectivity of lasers allows efficient schemes for practical isotope separation. Pilot plants exploring these new technologies are now in operation.

In view of the rapid progress in the field of laser technology in chemistry and the need for an authorative review of the development, leading experts in the field were invited to a one-day symposium in Stockholm on November 10, 1987. The symposium was arranged by the Royal Swedish Academy of Engineering Sciences (IVA) in coordination with the Swedish Board for Technical Development (STU). This Feature Issue of Applied Physics B consists of the invited papers delivered at the symposium.

The Guest Editors would like to thank all the authors for their excellent reviews. Special thanks are due to Dr. Helen Sheppard for her valuable help in the editing work. Dr. Helmut Lotsch and co-workers at Springer Verlag as usual performed very efficient work in the process of publishing these Proceedings.

Hans Medin
Royal Swedish Academy
of Engineering
Sciences
P. O. Box 5073
S-102 42 Stockholm

Sune Svanberg
Department of Physics
Lund Institute of
Technology
P. O. Box 118
S-221 00 Lund

Appl. Phys. B 46, 199-208 (1988)

Lasers for Chemical Applications

F. P. Schäfer

Max-Planck-Institut für biophysikalische Chemie, Abteilung Laserphysik,
D-3400 Göttingen, Fed. Rep. Germany

Received 29 February 1988/Accepted 9 March 1988

Abstract. The characteristics of lamps and lasers are contrasted from the viewpoint of a photochemist. The number of potential photochemical applications of lasers is shown to be vastly greater than that using conventional lamps. Several examples are given, which are only made feasible by the special properties of laser light.

PACS: 82.20. – w, 82.50. – m; 82.80. – d

When talking about chemical applications of lasers what first comes to one's mind is photochemistry, where lasers replace the time-honoured mercury-lamp of the photochemist. Perhaps this is why most chemists still think of a laser as an expensive lamp – and no more. I will later make a detailed comparison of lamps and lasers and actually demonstrate that laser photons can even be cheaper than lamp photons and one can use lasers in a multitude of sophisticated photochemical applications that one would never have dreamed to be possible with a lamp.

But, of course, there are many more chemical applications of lasers than just photochemical ones, e.g. analytical applications like trace analysis with infrared lasers, or coherent anti-Stokes Raman scattering and laser-induced fluorescence in combustion research, or laser mass spectroscopy. Then there are also kinetic or mechanistic investigations which have only become possible with ultrashort pulses from tunable lasers.

It is, of course, impossible to talk about all the different lasers that are useful in all these diverse applications, so I will only give a brief synopsis of the wavelength range covered by the different classes of lasers in the first figure and then concentrate on excimer lasers and dye lasers, which I believe cover more than 90% of all chemical applications.

The upper part of Fig. 1 contains the most important fixed-frequency lasers:

a) solid-state lasers, e.g. the ruby laser in the red end of the visible region and the Nd^{3+} glass or -YAG laser near 1 µm,

b) gas lasers, e.g. the He-Ne laser and the Ar^+- and Kr^+-ion lasers, with several lines throughout the visible,

c) excimer lasers, e.g. ArF (193 nm), KrF (248.5 nm), XeCl (308 nm), and F_2 (158 nm) in the near-UV and a so-called "X-ray laser" which was first operated in October 1984 at the Lawrence Livermore National Laboratory using a laser-generated Se-plasma emitting at 20 nm. At present, this XUV laser is still no more than a scientific curiosity but is listed here to remind one of the high hopes for future X-ray lasers which should operate down to wavelengths of a few Ångströms and will eventually open up a whole bonanza of important new applications.

In the infrared there are

d) the chemical lasers, e.g. the iodine-photo-dissociation laser at 1.35 µm and the HF laser with many vibronic lines near 2.5 µm,

e) the electrically excited molecular gas lasers, operating on many discrete vibronic lines, e.g. the CO laser around 5-6 µm and the CO_2 laser between 9 and 11 µm, and finally

f) the optically excited molecular gas lasers, e.g. water, formic acid, methanol and many others operating on thousands of pure rotational transitions in the far-infrared from about 30 µm extending into the millimetre region.

In the middle and lower part of Fig. 1 the wavelength range of the truly continuously tunable lasers is indicated. The low-power semiconductor lasers, which are important for analytical applications, operate from the red end of the visible to about 50 µm. Dye lasers

WAVELENGTH RANGE OF LASERS

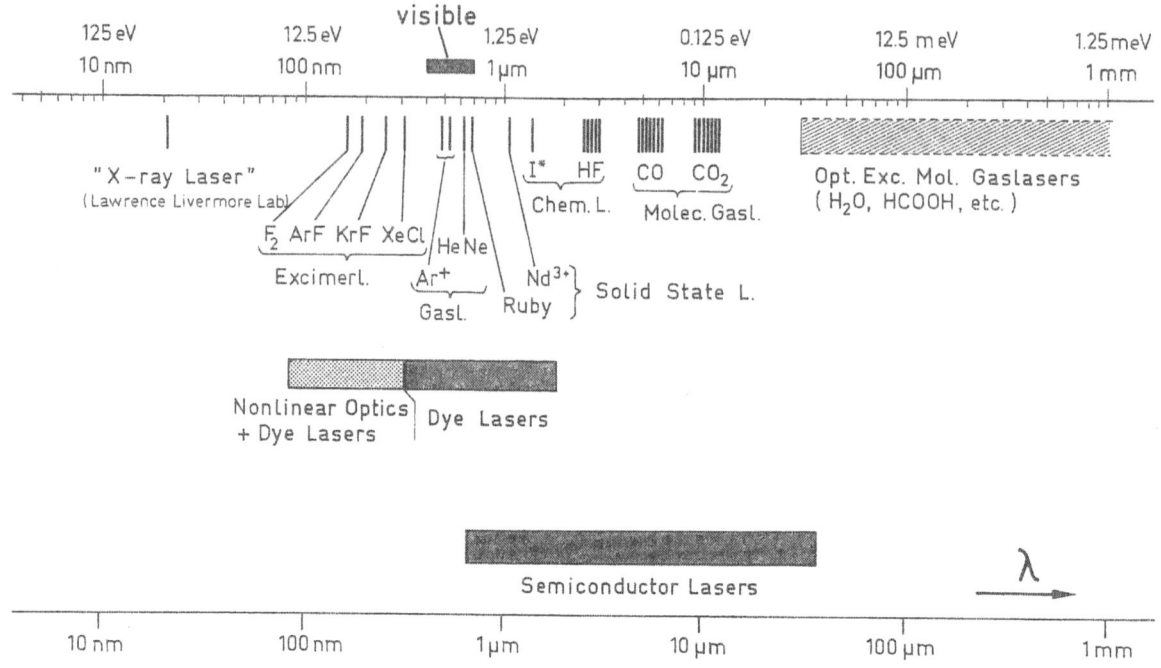

Fig. 1. Wavelength range of the most important lasers

presently operate in the range from 308.5 nm to 1.85 μm. Using nonlinear optical methods, e.g. frequency multiplication in crystals or gases, extends this range down to about 80 nm.

Comparison of Lamps and Lasers

Now let us compare the photochemically relevant properties of three often-used lamps and three typical lasers, namely an excimer laser, a dye laser, and a CO_2 laser (Fig. 2).

The 60 kW TL-doped mercury lamp is the largest lamp used in industry for a photochemical step in the high-volume production of nylon 6; the 700 W medium-pressure mercury lamp is a type most popular in the organic photochemist's laboratory; the high-pressure short-arc mercury lamp, HBO 200, used with a filter for singling out a spectral line or spectral region and a collimator to create a "parallel" beam is most popular with physical chemists and physicists.

Of the multitude of lasers three are chosen from different classes, which are relevant for photochemistry: a popular commercial KrF laser of 100 W average optical output power, a typical dye laser of 10 W average or continuous optical power, and a 10 kW CO_2 laser, which can be used in multiphoton-infrared photochemistry, where molecules absorb on the aver-

age about 20 photons, which results in the same molecular energy content as by absorption of a photon in the green.

The first line gives the quantum flux of these lamps and lasers, which determines ceteris paribus the yield per unit time in a photochemical reaction. Even in this simplest aspect lasers are at least on par, but often better than lamps. When intensity is considered (2nd line), lasers are seen to be orders of magnitude superior to lamps. The same is true for spectral width (3rd line), where lasers emit only on a single spectral line without any continuum: also for divergence (4th line), since all lasers naturally emit a "parallel" beam, as well as for pulse duration (5th line), where lasers are seen to be able to emit extremely short pulses. Finally, lasers can very simply be made to emit polarized light by inclusion of a Brewster window in the resonator, which in most cases is used anyhow.

Figure 3 compares the spectra of a lamp and some lasers. The upper part of the figure shows the lines of the above-mentioned 700 W mercury lamp, which are all emitted simultaneously. Selecting a single line or line pair with filters results in high losses of usually over 70%. On the other hand, the lines shown for the excimer laser in the lower part of the figure are only emitted one at a time, depending on the gas mixture used in the laser. The dye laser output shown as a

	Lamp			Laser		
	60 kW Hg-Tl at 535 nm	700 W Hg Medium Pressure (Original Hanau TQ 718) at 254 nm	HBO 200 + Collimator + Filter at 336 nm	KrF-Laser (Lambda-Physik EMG 302MSC) at 248.5 nm	Dye Laser 10 W at 590 nm	CO_2-Laser 10 kW at 10.6 μm
Quantum Flux $\left(\dfrac{\text{Einstein}}{h}\right)$	211	0.14	$<5 \cdot 10^{-4}$	0.73	0.18	$3.2 \cdot 10^3$ (162)
Intensity $\left(\dfrac{W}{cm^2}\right)$	3.5	0.01	$2.5 \cdot 10^{-3}$	$1 \cdot 10^7$	140	10^4
Spectral Width [nm]	ca. 20 + Lines	ca. 2 + Lines	ca. 2 + Continuum	1	$<10^{-3}$	<0.1
Divergence [°]	360	360	2.5	0.1x0.2	0.1	<0.1
Pulse Duration [s]	CW	CW	$CW->10^{-3}$	$2 \cdot 10^{-8}$	$CW-<10^{-10}$	<0.1
Polarization	unpolarized			Polarization >100:1 (with Brewster Window)		

Fig. 2. Comparison of some lamps with some lasers

shaded area is also emitting only one single line, which can be tuned over the range indicated. This single line emission of a laser vs. the multiline emission of a lamp is a great advantage of lasers applied to photochemistry.

E-beam-pumped excimer lasers that are capable of pulse energies of >15 kJ have been built in research institutions, and others are in the planning stage with MJ-pulse energies. However, the chemist is mostly interested in commercially available excimer lasers which are all discharge pumped. So Fig. 4 shows what one can expect from state-of-the-art excimer lasers.

Only the four most efficient excimers are shown. One should notice that, of course, not all specified values apply simultaneously, e.g., there is usually only a certain fraction of the maximum pulse energy available at the maximum repetition rate.

Let us have a look at the cost of one mole of photons from a XeCl laser to demonstrate that excimer lasers are by far cheaper than the high price of such an instrument seems to suggest. In Fig. 5 it is seen that the cost of photons is mainly determined by refurbishment costs at the present [1]. As the lifetime of energy storage capacitors, thyratrons (or other switches),

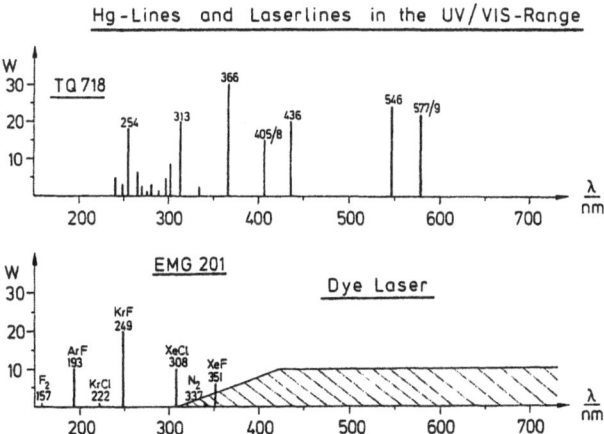

Fig. 3. Spectra of a mercury lamp used for preparative photochemistry and some lasers

Excimer Lasers - Standard Discharge Pumped

Excimer	ArF	KrF	XeCl	XeF	
λ	193	248	308	351	nm
Max. avg. power	60	120	100	50	W
Max. pulse energy	0.6	1	2	0.5	J
Max. pulse width	20	25	250	25	ns
Max. rep. rate	400	500	500	250	Hz
Min. Divergence (50% of energy)	<0.2 2	<0.2 2	<0.2 2	<0.2 2	mrad (oscill.-ampl.) mrad

Fig. 4. Characteristics of standard discharge-pumped excimer lasers

COSTS OF UV-LASER LIGHT

① DEPRECIATION OF THE LASER, ANNUAL

② OPERATING COSTS (LASER-GAS, ELECTRICITY, H_2O)

③ REFURBISHMENT COSTS

② ≪ ③   ① ≅ ③ DEPENDING ON USAGE OF CAPACITY

③ TYPICALLY $2 \cdot 10^4$ DM / $2 \cdot 10^8$ LASER SHOTS

↝ 10^{-4} DM / LASER SHOT

	Xe Cl 150 mJ	XeCl 1J
COSTS / MOLE (UV) (6·10²³ PHOTONS)	250 DM	37,50 DM
COSTS / kWh (UV)	2400 DM	360 DM

④ CREW COSTS

Fig. 5. Costs of UV laser light [1]

Dye Lasers – Standard Types

	Pulsed (Excimer- or Nd-l. pumped)	C.W. (Ar⁺- or Kr⁺-l. pumped)
λ-range	309 (189) – 1850 nm	≈ 400 – ≈ 1000 nm
Bandwidth	0.2 cm⁻¹	1 MHz (single freq.)
Peak power	> 20 MW (>1GW w. ampl.)	> 1 kW (cavity dumped)
Average power	> 10 W (Rh. 6G)	> 5 W (Rh. 6G)
Pulse width	< 1 ps - >100 ns	< 1 ps - ∞
Divergence	0.5 mrad	diffraction lim.

Fig. 6. Characteristics of typical dye lasers

windows, etc. will increase with further technical development, the cost of laser photons will also come down. Since a comparable calculation for the above-mentioned 700 W mercury lamp gives a cost of about 40 DM per mole UV [2], one sees that excimer lasers have already now reached break-even point with lamps as far as cost is concerned.

Figure 6 gives an overview of state-of-the-art dye lasers. The main division line here is between dye lasers for pulsed emission and those for continuous emission. As pulsed dye lasers only those pumped by excimer lasers or neodymium lasers are considered, since flashlamp-pumped dye lasers play only a minor rôle in some special applications. Cw-(continuous-wave) dye lasers are always pumped by either Ar⁺- or Kr⁺-ion lasers. (The recently reported first operation of a cw-dye laser with incoherent lamp pumping [3] at present cannot yet be assessed regarding its developmental potential as a photochemically interesting light source.) For special applications in laser isotope separation (v.i.) pulsed dye lasers of average power > 300 W pumped by copper-vapour lasers at a multi-kHz repetition rate have been developed.

Laser Applications in Chemistry

After these introductory remarks on the interesting properties of lasers in contrast to lamps, one might want to know what new and unique applications in chemistry are made possible by lasers. If we want to bring the multitude of new applications into a certain systematic order, we should apply the morphological method of the late Swiss astronomer Zwicky [4] and construct a morphological box, as shown in Fig. 7.

Here all possible combinations of the five properties of lasers, which are important for the chemist, namely intensity, monochromaticity, collimation, pulse length, and polarization, are listed. Coherence, another property of lasers, which is important for many physical applications is not considered in this context. Thus, all applications which make use of only one of these properties of lasers are listed in the first five fields in the box, those applications which make use of two of these properties simultaneously are filed in the next ten fields, and so on. One finds that a complete listing contains 31 classes of laser applications in chemistry, which could hardly be realized with lamps.

Morphological Box for Laser Applications

Types of Laser Applications	Important Properties of Lasers and Combinations Thereof
5	Intensity I Monochromaticity M Collimation C Pulselength T Polarization P
10	I+M I+C I+T I+P M+C M+T M+P C+T C+P T+P
10	I+M+C I+M+T I+M+P I+C+T I+C+P I+T+P M+C+T M+C+P M+T+P C+T+P
5	I+M+C+T I+M+C+P I+M+T+P I+C+T+P M+C+T+P
1	I+M+C+T+P

Fig. 7. Morphological box for laser applications

Absorption

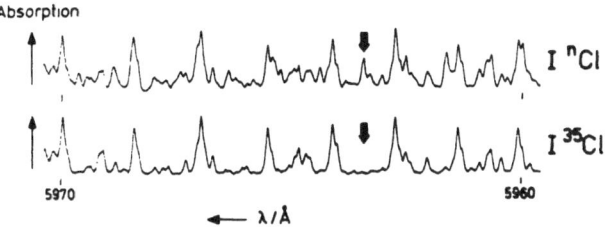

Fig. 8. A 1 nm section of the spectrum of ICl; upper spectrum for a sample with natural isotopic composition, lower spectrum for a sample which is highly enriched in ^{37}Cl

Selective Photoaddition

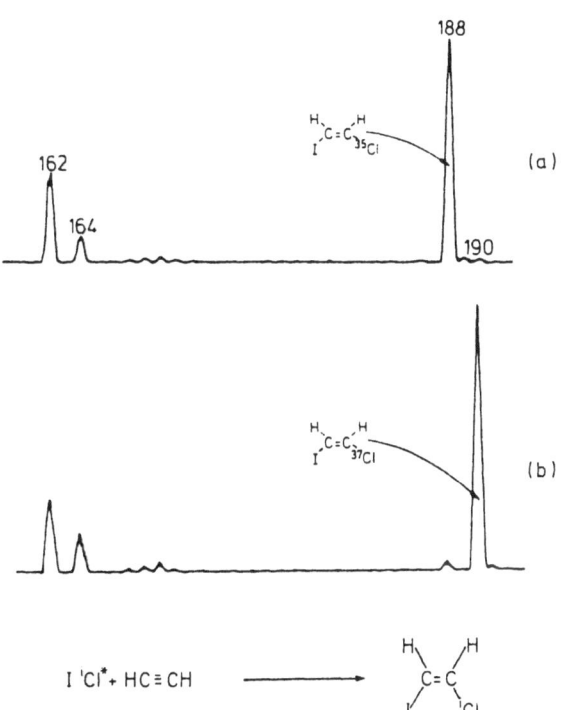

Fig. 9. Mass spectra of the educts and products of the photoreaction shown on the bottom line, starting with a sample of natural isotopic composition. Upper spectrum: irradiation at a line of $I^{35}Cl$, lower spectrum: irradiation at a line of $I^{37}Cl$

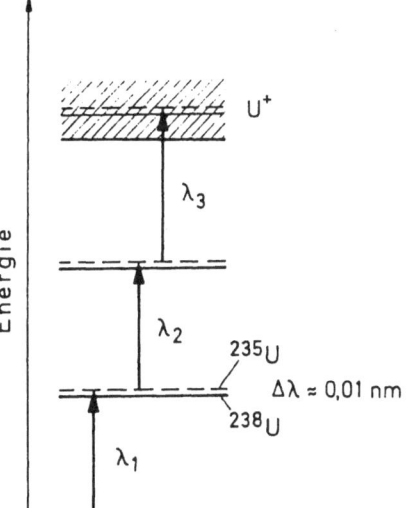

Fig. 10. Schematic demonstration of the AVLIS principle [6]

caused by the content of $I^{37}Cl$ in the sample. This is most clearly seen at the wavelength marked by an arrow, where $I^{35}Cl$ does not absorb at all. Irradiating a sample of natural isotope abundance at this wavelength with a properly tuned dye laser, will only excite $I^{37}Cl$ molecules. If we now add a reactant, which only reacts with molecules in the excited state, the reaction will be highly isotopically selective. The result of such a reaction, in this case a photoaddition of acetylene to iodine chloride, is shown in Fig. 9.

Two mass spectra of partly reacted mixtures are shown. The peaks at masses 162 and 164 belong to $I^{35}Cl$ and $I^{37}Cl$, respectively, those at 188 and 190 to the corresponding reaction products. For the upper trace an irradiation wavelength was chosen where only $I^{35}Cl$ absorbs, while for the lower trace the other isotopomer was chosen. As can be seen here, a high enrichment of the desired isotope is obtained. Depending on the reaction conditions we found enrichment factors of over 100 in a single step.

A similar scheme for laser-isotope separation, which is much more important, however, than the one just discussed, is the Atomic Vapour Laser Isotope Separation scheme (AVLIS for short) for the enrichment of uranium. The schematic diagram of Fig. 10 shows only a few energy levels from the huge number actually found in uranium vapour. The isotope splitting between ^{235}U and ^{238}U, which is typically 0.01 nm for many transitions, are indicated [6]. Since dye lasers can easily be tuned to the ^{235}U line, selective excitation of this rare isotope is achieved. To enhance the selectivity, a second dye laser at a properly chosen second wavelength transfers the population of the first excited level to a second level, and finally a third dye laser at a third wavelength brings the atoms into the

The three fields into which the following examples belong, are underlined in the morphological box.

As the first example of an application, where only one property of the laser, namely the monochromaticity is used (together, in this case, with the tunability), let us consider the laser-isotope separation of the chlorine isotopes that we did several years ago [5]. In Fig. 8 a part of the spectra of ICl is shown, recorded with a dye laser. The upper spectrum is obtained with a sample of natural isotope abundance of chlorine, the lower one with a ^{35}Cl-enriched sample. The bands in the upper spectrum which are missing in the lower one, are

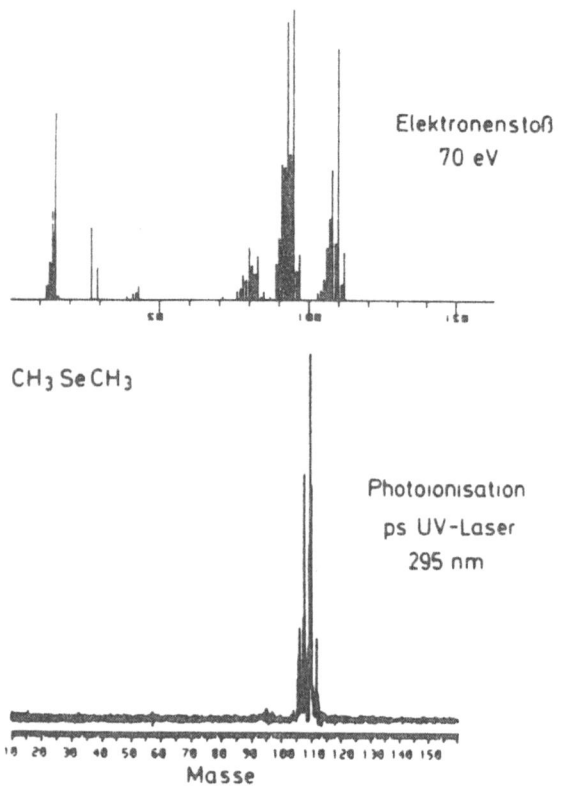

Fig. 11. Mass spectra of CH_3SeCH_3. Upper trace: ionization by electron beam; lower trace: ionization by a ps-UV laser [7]

make an optical absorption path length of several metres absolutely mandatory.

The next example is one from laser mass spectrometry and makes use, in addition to the properties just mentioned, also of the capability of the laser to produce ultrashort pulses. It is taken from [7] and is illustrated in Fig. 11. This example gives the two mass spectra of a metal-organic compound, CH_3SeCH_3, useful for laser-chemical vapour deposition (LCVD). The upper trace is taken in the conventional way, with an electron gun of 70 eV ionizing the molecules. One can notice the usual pattern of the many different fragments created by the electron collisions. A similar pattern is also obtained if one irradiates the molecules by a strongly focussed dye laser beam at a wavelength that is absorbed by the molecules. Multiphoton absorption first ionizes the molecules and further multiphoton absorption by the ions causes dissociation into many smaller fragments during the usual irradiation pulses of 10–20 ns pulse length. There is, however, one advantage of laser ionization over electron beam ionization in the case of mixtures of several molecular species: choosing a wavelength where only the species of interest absorbs, one selectively ionizes only this species without perturbation by the other species present.

A much more important advantage, however, is obtained when using a picosecond dye laser pulse, as shown in the lower trace of Fig. 11. Here only the mother peaks of CH_3SeCH_3 appear (6 peaks of the isotopomers with the known natural abundance ratios). This is so, because the ion created in the ps pulse cannot be dissociated any more since there are no more photons available, in contrast to ns pulse irradiation.

Of principal importance are all multi-photon processes, since the population density of some excited states is intensity dependent. The intensity which is necessary to keep one half of the molecules in some excited state is given in Fig. 12. From triplets in

ionization continuum. The $^{235}U^+$ ions are deflected by an electric field and collected at the electrode. More details of this AVLIS process and the huge demonstration plant under construction by the Lawrence Livermore National Laboratory will be reported in the paper by J. A. Paisner later in this issue.

This is an example where the monochromaticity, the intensity and the collimation of the laser are absolutely essential, since the population of the excited states is intensity dependent and the small absorption cross-sections and small densities of uranium vapour

Intensity $I_{1/2}$, to hold 1/2 of all molecules in the excited state (with lifetime τ):

$$I_{1/2} = \frac{N_L \cdot h \cdot \nu}{1000 \cdot \ln 10 \cdot \varepsilon \cdot \tau} \quad \text{(optically thin solution)}$$

Numerical example for $\varepsilon = 10^4$ $l \cdot M^{-1} \cdot cm^{-1}$, $\lambda = 590$ nm

τ	Realization	$I_{1/2}$		Laser
10 μs	Triplet, degassed solution	880	W/cm^2	cw, weakly focussed
100 ns	Triplet, air saturated solution	88	kW/cm^2	cw, strongly focussed
1 ns	Singlet	8,8	MW/cm^2	Pulsed laser, 10-ns-pulse
1 ps	Vibrational level	8,8	GW/cm^2	Ultrashort laser pulse (10 ps)

Fig. 12. Tabulation of intensities needed to hold half of the irradiated molecules in the excited state with lifetime τ

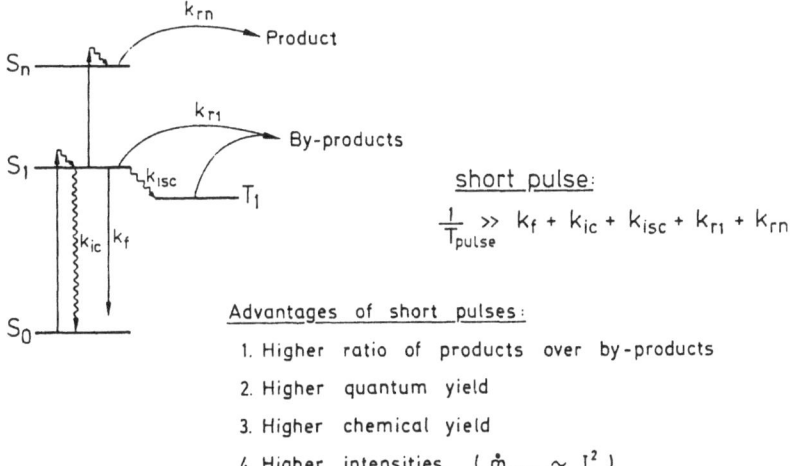

short pulse:

$$\frac{1}{T_{pulse}} \gg k_f + k_{ic} + k_{isc} + k_{r1} + k_{rn}$$

Advantages of short pulses:

1. Higher ratio of products over by-products

2. Higher quantum yield

3. Higher chemical yield

4. Higher intensities ($\dot{m}_{Product} \sim I^2$)

Fig. 13. Advantages of using short pulses in photochemistry

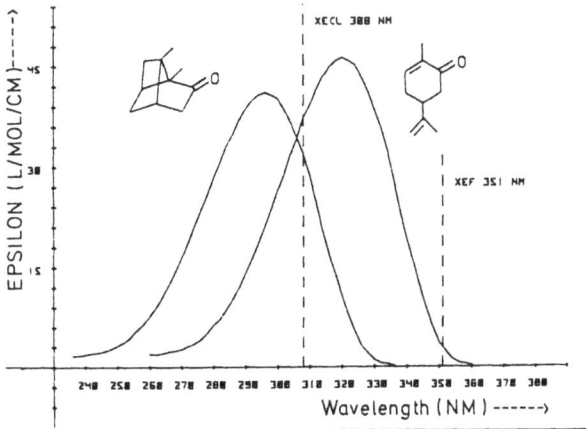

Fig. 14. Spectra of carvone and carvone camphor

degassed room temperature solutions with an intensity $I_{1/2} \approx 1$ kW/cm^2 to vibrational level excitation the intensity rises to $I_{1/2} \approx 10$ GW/cm because of the shorter lifetime.

The higher the intensity which one needs to keep one half of the molecules in the excited states, the shorter the laser pulse must be for a given pulse energy. But short pulses actually have some advantages for reactions from higher excited states, as shown in Fig. 13. The relative importance of these advantages depends, of course, on the specific reaction and substances involved.

I would like to give two examples of reactions from higher excited states. The first is the famous example of the isomerization of carvone to carvone camphor [8] and the molecular formulae and the spectra are shown in Fig. 14. The dashed lines indicate the wavelengths of the two excimer lasers which can be used for this photoreaction. If XeF is used, only the educt absorbs and reacts, if, however, XeCl is used, also the product

absorbs, which results in a consecutive reaction leading to an (usually unwanted) ester in alcoholic solution at room temperature.

The result of the reaction executed 1) with a mercury lamp, 2) with a XeCl laser, 3) with a XeF laser is shown in Fig. 15. The fact that the quantum efficiency increases from near zero (in the case of the lamp with low intensity) to 7% at high laser intensity is a clear indication that this is a two-photon reaction involving higher excited states. The resulting products are practically pure, without any of the brown, tarry byproducts obtained with lamp irradiation [9].

The second example was studied earlier with mercury lamps and found to be a 2-photon process [10], as shown in Fig. 16. While at room temperature in liquid solution no reaction occurs with even the strongest lamp irradiation, in organic glass at 77 K the reaction only occurs when irradiation at *both* mercury lines, at 313 and 404 nm, is used simultaneously. It was assumed that the short wavelength first populates the lowest triplet state, while the longer wavelength transfers the molecules into a higher lying reactive triplet.

We have irradiated this molecule and two similar ones with a laser at room temperature and found that we could obtain a second product, as seen in Fig. 17. We supposed that the second reaction path was via a higher lying reactive singlet state, since the high-intensity laser can populate the lowest lying singlet state notwithstanding its short lifetime, as discussed above, so that a transition to the higher lying reactive singlet by the absorption of a second laser photon becomes highly probable, proportional to the laser intensity.

We investigated this in more detail [11] with a combination of one excimer laser plus one dye laser and found some interesting results, as shown in Fig. 18. Simultaneous irradiation at 308 and 650 nm led main-

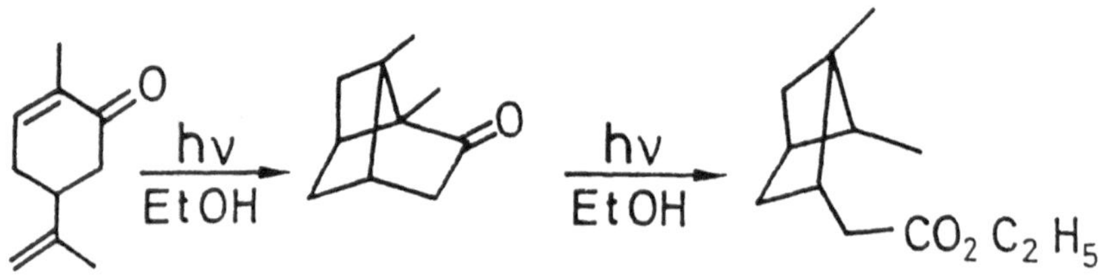

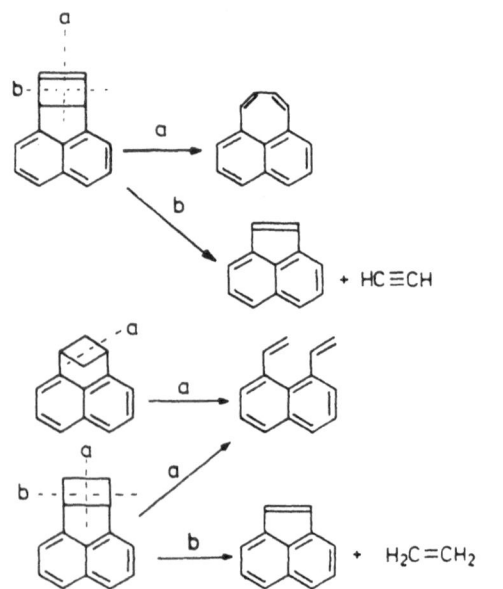

The carvone reaction scheme (C → K → E) at top:

C   →hv/EtOH→   K   →hv/EtOH→   E (with CO₂C₂H₅)

	Quantum Efficiency	Ratio E/K
Hg-Lamp (313 nm)	$<3 \cdot 10^{-3}$	0.54 + byproducts
Laser (XeCl, 308 nm)	increases with I_0 to 7 %	decreases with I_0 to 0.15
Laser (XeF, 351 nm)	increases with I_0 to 7 %	0.07

Fig. 15. Results for the carvone-carvone camphor rearrangement done with lamp and lasers [8]

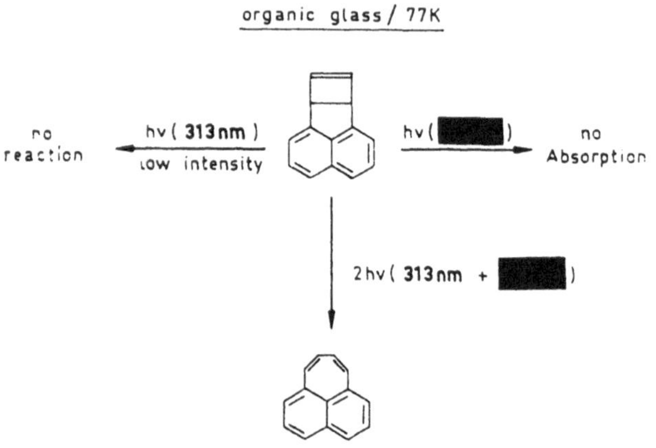

organic glass / 77K

no reaction   ←hv (313nm)/low intensity—   [structure]   —hv (███)→   no Absorption

2hv (313nm + ███)

Fig. 16. Two-photon reaction with lamp irradiation in an organic glass at 77 K [10]

Fig. 17. Possible reaction pathways for some aromatic-substituted cyclobutanes and cyclobutene [11]

ly to compound 3, while at 308 nm plus a 160 ns delayed pulse at 440 nm led almost exclusively to compound 2, as in the organic glass with lamp irradiation. The ratio 2/3 was always intensity dependent. From these facts it is clear that the first reaction path is via a higher lying reactive singlet S_n, while the second one is via a higher lying reactive triplet T_n.

So far I have only shown photochemical reactions, which we followed with conventional techniques. But if one wants to learn about the mechanism of the reaction, one can often do so by studying the kinetics on a time scale that is adequate for the reaction under study. This usually means ps- or sub-ps pulses, if the primary steps are to be resolved.

Normally this means a very complex experimental set-up with several lasers and a lot of electronic and electrooptic equipment that needs at least one physicist to keep it running. We wanted to develop a much simpler device for pump-probe experiments, so simple, if possible, that any chemist or biologist can switch on the system in the morning and immediately start to work with it, not on it. Although we have already gone a long way in this direction, as will be seen immediately, we probably need one more year to reach this goal and probably another year until it becomes commercially available.

Figure 19 shows a block diagram of our hybrid excimer-laser/dye laser system for the generation of fs pulses in the UV and visible. It consists of an excimer laser with two discharge channels triggered simultaneously (Lambda Physik, EMG 150). One channel is surrounded by a standard laser resonator and its output at 308 nm is used to pump a number of dye cells forming a ps dye laser system and another set of dye cells arranged to form an amplifier chain for ultrashort pulses at 616 nm. The output of the fs dye laser system has a wavelength of about 380 nm and a pulse length of

Liquid Solution at Room Temperature

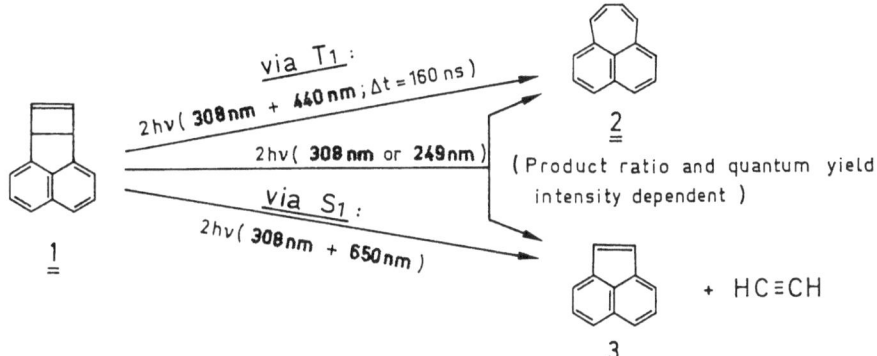

Fig. 18. Reaction pathways with laser irradiation [11]

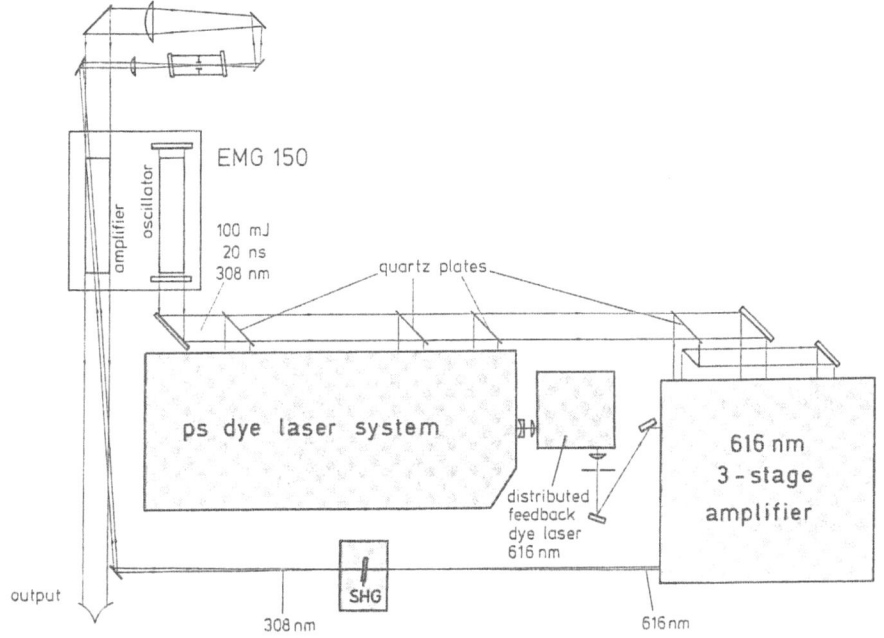

Fig. 19. Arrangement for the generation of ps excimer-laser pulses [12]

8 ps and is used to pump a special dye laser ("distributed feedback dye laser") which creates subpicosecond pulses at 616 nm. These pulses are amplified in the above-mentioned dye laser amplifier chain to high peak powers (several GW) and are then frequency-doubled in a potassium dihydrogen phosphate (KDP) crystal to give a few μJ at 308 nm. These pulses are then sent through the second amplifier channel of the EMG 150, where they are amplified to several mJ, while the pulse width is reduced below 300 fs [12].

One simple experiment, which demonstrates the temporal resolution of this system is shown in Fig. 20. The 308 nm pulse and a fraction of the 616 nm pulse are sent collinearly to a dichroic mirror, where they are separated and sent separately over two delay lines to a dye cell, which contains a bifluorophoric dye (v.i.). The blue absorbing moiety of the dye molecules is

excited by the 308 nm pulse and within a very short time transfers the excitation energy to the longer wavelength absorbing moiety. Since this latter moiety emits fluorescence peaking near 616 nm, the red pulse will experience amplification, when it arrives at the dye cell at or shortly after the time of the 308 nm pump pulse. By changing the delay of the red pulse from shot to shot, one can follow the rise and fall time of the gain. The difference between this time function and the one calculated for the case of direct excitation is the time needed for the energy transfer from the absorbing to the emitting dye moiety [13].

The result of the intramolecular energy-transfer measurements can be seen in Fig. 21. For the molecule shown in the inset the rise of the gain follows the curve connecting the measurement points which coincides with the curve calculated for a transfer time

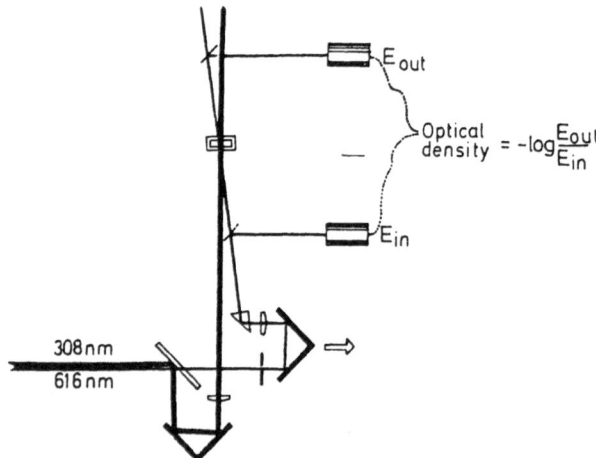

Fig. 20. Experimental pump-probe arrangement [13]

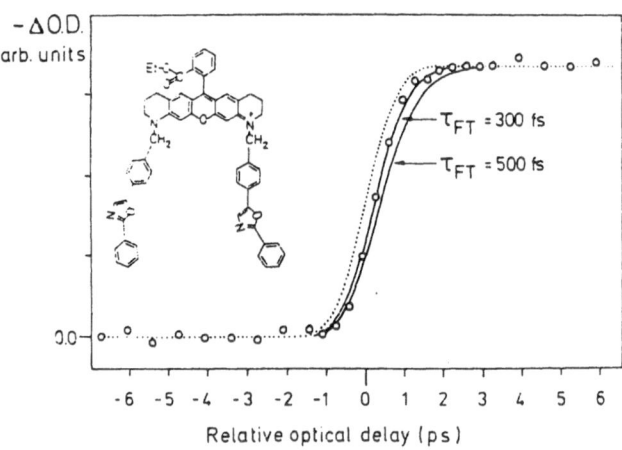

Fig. 21. Measurement of the intramolecular energy transfer in the bichromophoric molecule shown in the inset [13]

$\tau_{FT} = 300$ fs. The other solid line is the calculated curve for $\tau_{FT} = 500$ fs, while the dotted line is calculated for zero transfer time.

This brief description of a laser system for kinetic measurements, which is capable of following even the fastest primary processes in photochemical reactions, is the end of this review in which I endeavoured to give an overview of the state of the art in lasers for chemical applications and I hope to have outlined several perspectives on how to exploit the marvellous properties of lasers for chemical purposes.

References

1. Lambda Physik, Göttingen: Private communication
2. F.P. Schäfer: Unpublished
3. E. Thiel, C. Zander, K.H. Drexhage: Opt. Commun. **60**, 396 (1986)
4. F. Zwicky: *Morphological Astronomy* (Springer, Heidelberg 1957)
5. M. Stuke, F.P. Schäfer: Chem. Phys. Lett. **48**, 271 (1977)
6. G. Meyer-Kretschmer, H. Jetter: Naturwissenschaften **70**, 7 (1983)
7. R. Larciprete, M. Stuke: J. Crystal Growth **77**, 235 (1986)
8. U. Brackmann, F.P. Schäfer: Chem. Phys. Lett. **87**, 579 (1982)
9. G. Büchi, I.M. Goldmann: J. Am. Chem. Soc. **79**, 4741 (1957)
10. J. Meinwald, G. Samuelson, M. Ikeda: J. Am. Chem. Soc. **92**, 7604 (1970)
11. D. Plaas, F.P. Schäfer: Chem. Phys. Lett. **131**, 528 (1986)
12. S. Szatmári, B. Rácz, F.P. Schäfer: Opt. Commun. **62**, 271 (1987)
13. N.P. Ernsting, M. Kaschke, J. Kleinschmidt, K.H. Drexhage, V. Huth: Chem. Phys. (submitted)

Appl. Phys. B 46, 209–220 (1988)

The Impact of Several Atomic and Molecular Laser Spectroscopic Techniques for Chemical Analysis

N. Omenetto

Joint Research Centre, Chemistry Division, I-21020 Ispra (Varese), Italy

Received 11 March 1988/Accepted 14 March 1988

Abstract. Several laser-based methods, namely laser induced fluorescence, laser enhanced ionisation and thermal lensing spectrophotometry are discussed with respect to their capabilities of approaching the extremely high detection sensitivity which is nowadays required in many fields of application, notably in high purity materials, in biomedicine and in the nuclear industry. The discussion is restricted to atomisers operated at atmospheric pressure, i.e., combustion flames, plasmas and graphite furnaces. It is shown that the analytical limit of detection can be in the range of femtograms and that double-resonance excitation possesses significant advantages over single-resonance excitation, both in terms of signal-to-noise ratio and spectral selectivity. In addition, the combination of the fluorescence and ionisation techniques represents a remarkable diagnostic tool. In the nuclear field, the suitability of the technique of thermal lensing for the direct determination and chemical speciation of very low levels of uranium in water is discussed.

PACS: 82.80, 32.00, 07.65

Chemical analysis is intuitively associated with the concept of detecting, on a qualitative as well as quantitative basis, virtually any atom or molecule present in any given sample, irrespective of its chemical composition and physical state. To accomplish this in practice, one would need a perfect complete analytical procedure, capable first of all of efficiently decomposing the sample into its constituents, which can then be selectively addressed and individually detected. However far from reality such an ideal technique might be, a continuous effort has been devoted to the development of extremely sensitive methods of analysis and among these, the atomic and molecular spectroscopic methods have played a prominent role.

A panoramic view of the detection limits offered by several atomic spectrometric techniques [1–3] indicates that values of the order of 10 pg/ml can effectively be obtained for a few elements by the techniques of atomic absorption and atomic fluorescence with electrothermal atomisation or atomic/ionic emission with the inductively coupled plasma source, especially if combined with a mass spectrometer. If one assumes that the sample (1 g) had to be dissolved in 100 ml of solution which was subsequently analysed, the above limit corresponds to the detection of 1 part in 10^9 in the original sample. Although such detection power is perfectly adequate for many analytical applications, there is a need for still higher sensitivity in important fields such as the nuclear industry, the electronic and semiconductor industry and the biomedical industry. In many cases, the average sensitivity needed is as high as 1 part in 10^{13} or 0.1 pg/g. By the same dissolution procedure mentioned above, this would require a concentrational limit of detection of 1 fg/ml, which is orders of magnitude below that reached by conventional spectrometric methods. This amply justifies the development of other methods, irrespective of their complexity and cost. Laser sources, in particular tunable lasers, have set up the framework upon which these methods have developed [4–9].

The use of lasers in atomic and molecular physics has had a profound impact, opening up a new class of spectroscopic techniques which allow a detailed investigation of the structure of atoms and molecules [10]. In analytical spectrochemistry, laser-based methods have been (and still are) developed at an impressive rate. This is essentially due to the peculiar attributes of lasers which not only allow the improvement of already well established methods (e.g., absorption, fluorescence, Raman and optoacoustic methods) by increasing their sensitivity and selectivity, but also make possible a great variety of new interaction processes and analytical methodologies (e.g., multi-photon excitation, intracavity absorption, laser enhanced ionisation and resonance ionisation, remote sensing methods and thermo-optic methods).

In this paper, the discussion is restricted to the techniques Laser Induced Fluorescence (LIF), Laser Enhanced Ionisation (LEI), and Thermal Lensing Spectrophotometry (TLS). The former two methods have been used in combination with atmospheric pressure atomisers (such as flames, plasmas, and graphite furnaces) as well as with atomisers operated under vacuum. The TLS method will be discussed with regard to the direct speciation of uranium in water solutions as well as to trace analysis of metals extracted in an organic solvent.

1. Laser-Induced Fluorescence and Laser-Enhanced Ionisation

Optical and *electrical* techniques based upon laser excitation have been considered capable of extremely high sensitivities for atomic detection, down to the single atom level [11–14]. In the former case, we are dealing with resonance and non-resonance fluorescence processes, while in the latter case one measures the change in the current flowing between biased electrodes placed in the atom reservoir (optogalvanic effect) [15]. There seems to be a semantic distinction (see later) between LEI and Resonance Ionisation Spectroscopy (RIS). LEI involves endoergic collisions to bridge the gap between the ionisation energy of the atom and the excitation energy provided by one or more excitation steps involving bound transitions, while RIS refer to the direct photo-ionisation of the laser populated state and therefore can be applied in collision-free environments. The distinction between LEI and RIS seems therefore related to the atom reservoir. While in atomic beams and atomisers operated at low pressures RIS is the obvious process, in atmospheric pressure atomisers and at elevated temperatures, both LEI and RIS can be effective, and it becomes difficult to ascertain which of the two mechanisms predominates.

The atomic fluorescence technique, together with its applications, has been exhaustively reviewed in [4,

5, 8, 9], while RIS is treated in [8, 12–14]. A comprehensive coverage of LEI, mainly in combustion flames, is found in [5, 9] and especially in [16].

1.1. General Considerations

A simplified picture of several excitation schemes used for LIF and LEI in flames and other atomisers is shown in Fig. 1. Here, the symbol B refers to the Einstein coefficient for stimulated absorption and emission while A is the coefficient for spontaneous emission. The $B\varrho_\lambda$'s represent the stimulated absorption (emission) rates and hold for a *broad-band* radiation field, of spectral energy density ϱ_λ, which is assumed to be spatially uniform within the excitation volume and temporally constant during the interaction with the atomic system. The symbol k indicates the collisional rate coefficients, and i stands for the ionisation continuum, with ionisation energy E_i.

Referring to this figure, the following LIF and LEI processes have been used in the literature [4, 5, 8, 9, 15–23] for analytical and diagnostic purposes:

1.1.1 Single-Resonance Fluorescence. In this case (which can also be called *single-step* or *single-colour* fluorescence), atoms are excited from level 1 (usually the ground state) to level 2, and resonance fluorescence (A_{21}) or direct-line fluorescence ($A_{22'}$) are measured. The advantage of using the non-resonance scheme ($1 \rightarrow 2$ excitation, $2 \rightarrow 2'$ detection) lies in the fact that the fluorescence monochromator isolates the true fluorescence signal from the spurious scattering occurring at the same laser excitation frequency. This scattering, which is essentially of the Mie type and is due to unvaporised particles formed when a sample solution with a high content of dissolved solids is aspirated into the flame, is very detrimental for the analysis and in several cases negates the use of the resonance fluorescence process. In addition to such scattering, in a fluorescence set-up it is difficult to completely prevent some laser photons (reflected from the optical components or even from atmospheric dust) from entering into the detection system. As a result, the non-resonance scheme is always preferred.

The spectral selectivity of this scheme is expected to be high, since it relies both upon the laser source, which is sharply tuned to a particular atomic absorption transition, and upon the fluorescence monochromator, which isolates that particular fluorescence transition in the (already simple) fluorescence spectrum emitted by the excited atom.

1.1.2 Double-Resonance Fluorescence. In this case (which can again be called *two-step* or *two-colour* fluorescence), two laser beams are simultaneously directed, in spatial and temporal coincidence, into the atomiser. When tuned to transitions $1\rightarrow2$ and $2\rightarrow3$, ground state atoms can be efficiently pumped into level 3 from which fluorescence can be monitored at either transition $3\rightarrow2$ (A_{32}) or $3\rightarrow3'$ $(A_{33'})$. The same considerations given above for the scattering problems are also valid here. However, the double-resonance scheme offers the unique advantage of correcting the scattering signal which affects the resonance fluorescence measurement at λ_{32}. In fact, since the signal monitored at this wavelength would essentially be zero when the first laser (tuned at λ_{12}) is not present, because of the negligible thermal population of level 2, any residual signal observed under these conditions will only be due to scattering. Of course, scatter *noise* will not be eliminated and therefore a degradation of the signal-to-noise ratio will occur.

When the second laser transition starts from the same level reached by the first laser (e.g., level 2 in the figure) one can speak of *connected* double-resonance excitation. However, in atmospheric pressure atomisers, collisional coupling between neighbouring levels is very effective. As a result, *disconnected* double-resonance excitation can also be utilized. This would imply exciting the atoms from level 1 to level 2 and then from level 2' to level 3. As indicated in the figure, level 2' is populated by radiation $(A_{22'})$ and by collisions $(k_{22'})$.

Compared to the single-resonance scheme, the spectral selectivity of the double-resonance approach will be extremely high. Indeed, the excitation selectivity will now be due to the *product* of the individual selectivities pertinent to each absorption transition, and, in addition, the monitored fluorescence transition can still be selected by the monochromator.

1.1.3 Single-Resonance Excitation, Collisional Ionisation. In this process, the laser is used to efficiently pump the atoms in level 2 from which thermal ionisation by collisions in the flame proceeds at a much higher rate compared with that of the ground state. k_{2i} expresses the collisional ionisation rate coefficient and depends upon the flame composition and temperature. Unlike the case of fluorescence, the spectral selectivity of the process here has to rely solely on the excitation step, i.e., on the spectral bandwidth of the laser.

1.1.4 Double-Resonance Excitation, Collisional Ionisation. As in the case of double-resonance fluorescence, two laser beams now populate level 3 (either from level 2 or 2') from which collisional ionisation proceeds very rapidly, the more so if a Rydberg state is reached with the second laser step. The spectral selectivity of the overall process is now much improved in comparison with the previous one.

1.1.5 Two-Step Ionisation. The process described in Sects. 1.1.3 and 1.1.4 might not be the only ones

responsible for the ionisation mechanism. In fact, if the energy of the laser photon is greater than half the energy required to ionise the atoms directly from the ground state (case in Sect. 1.1.3), or greater than the difference $E_i - E_2$ (case in Sect. 1.1.4), direct photoionisation by the absorption of a second laser photon at the same frequency might occur. This process is more likely to be observed in the case of double-resonance excitation since the energy of the photon at λ_{23} is often greater than the difference between the ionisation continuum and the excitation energy of level 2. Such photoionisation processes are the natural extension of RIS to flames.

Two-step photoionisation (or dual-laser ionisation) can of course be achieved by using one laser, tuned to a bound transition, to excite atoms, and another laser to bring the excited atoms to the continuum (*non-resonant photoionisation*) or to a narrow autoionisation state (*resonant photoionisation*). The former scheme has been applied in flames and with pulsed laser excitation by using the nitrogen or excimer lasers as the photoionising beams [19]. In this case, however, because of the collisional coupling between levels, it is not simple to unambiguously define terms such as *energy defect* or *energy overshoot*, which refer to the difference between the ionisation energy and the energy of the photoionising laser photon.

Multi-photon processes (not indicated in Fig. 1) and involving virtual levels are also possible and have been observed for analyte atoms in flames, in some cases with remarkable analytical sensitivity. It should be noted, however, that multiphoton ionisation signals may be due to the sample matrix constituents as well, and in addition, to native flame constituents (e.g., NO), giving rise to undesirable background effects.

1.1.6 Analytic Signal Expressions. When the rate equations for the populations of the levels shown in Fig. 1 are solved, with the assumption that the transitions are optically saturated, and for a rectangular laser pulse of duration Δt_l, which is too short for taking into account recombination effects, we can write the following expressions for the total number of ions formed during the interaction

$$n_i = n_T \left\{ 1 - \exp\left[-\left(\frac{g_u}{\Sigma_g}\right) R_{ui}\Delta t_l \right] \right\}, \tag{1}$$

$$n_i = \frac{n_T R_{ui}}{R_{ui} + R_{um}}$$
$$\times \left\{ 1 - \exp\left[-\left(\frac{g_u}{\Sigma_g}\right) (R_{ui} + R_{um})\Delta t_l \right] \right\}. \tag{2}$$

In Eq. (1), n_T indicates the total number of atoms present in the laser excitation volume, g_u is the

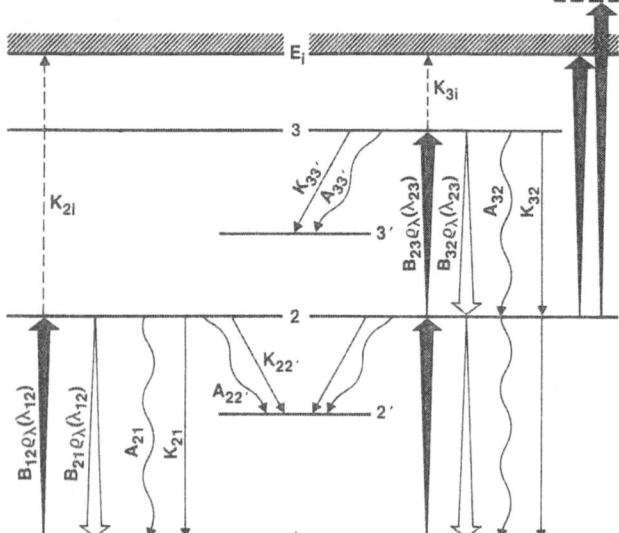

Fig. 1. Simplified energy level scheme describing several excitation-ionisation processes. Stimulated absorption and emission are indicated by thick arrows pointing upwards and downwards, respectively. Fluorescence transitions are shown by wavy arrows while collisional transitions are represented by straight arrows. Direct photoionisation, where a second laser photon reaches the ionisation continuum or an autoionising level above it, is also indicated. Refer to the text for the explanations of the symbols used. (From [17], with permission)

statistical weight of the upper level of the transition, Σg indicates the summation of the statistical weights for all levels involved in the interaction and R_{ui} is the total ionisation rate out of the upper level u. For example, $R_{ui} = k_{2i}$ or k_{3i} and $\Sigma g = (g_1 + g_2)$ or $(g_1 + g_2 + g_3)$ in the cases described in Sects. 1.1.3 and 1.1.4, respectively. It is clear that a unity ionisation yield in flames is obtained if the product $R_{ui}\Delta t_l$ is much greater than unity.

Equation 2 takes into account the existence of a metastable level, m, (e.g., level 2′ in Fig. 1) and shows that, in order to achieve efficient ionisation, R_{ui} must be greater than R_{um}, i.e., greater than the loss rate in the metastable level.

For the fluorescence counterpart, if we compare the double-resonance (DR) with the single-resonance (SR) schemes shown in Fig. 1, the ratio of the maximal saturated fluorescence radiances at λ_{32} and λ_{21}, under the assumption of steady-state conditions and negligible ionisation losses, is given by the expression

$$\frac{(B_f)_{DR}}{(B_f)_{SR}} = \left(\frac{A_{32}}{A_{21}}\right)\left(\frac{\lambda_{21}}{\lambda_{32}}\right)$$
$$\times \left(\frac{g_3}{g_2}\right)\left(\frac{g_1 + g_2}{g_1 + g_2 + g_3}\right) \qquad (3)$$

which shows that no loss in sensitivity is expected for the DR scheme, provided that A_{32} is as large as A_{21} and the ratios in parentheses are close to unity.

As is clear from the preceeding discussion, however, the ionisation losses cannot be considered negligible a priori, expecially in the double-resonance excitation scheme. It can be shown that the integrated fluorescence radiance pulse, is given by the expression

$$B_f = C \int_0^{\Delta t_l} n_u(t)\,dt$$
$$= \frac{Cn_T}{R_{ui}}\left\{1 - \exp\left[-\left(\frac{g_u}{\Sigma_g}\right)R_{ui}\Delta t_l\right]\right\}, \qquad (4)$$

where C is a constant of proportionality, including the spontaneous emission coefficient as well as geometrical and instrumental parameters. For $R_{ui}\Delta t_l \ll 1$, the maximum (saturated) fluorescence signal results. For $R_{ui}\Delta t_l \gg 1$, the fluorescence plateau will be much lower and such a decrease (fluorescence dip) may be exploited for diagnostic purposes (see later). Equation (4) shows that the double-resonance atomic fluorescence technique in flames might suffer from severe ionisation losses. In this case, a much better approach is the use of double-resonance *ionic* fluorescence in a high-temperature argon plasma (where natural ionisation is very pronounced) since in this case ion losses will be negligible, due to the large value of the ionisation potential of the ion.

1.1.7 Noise Considerations. In LIF experiments with flames, plasmas, and furnace atomisers, the limiting noises in the system, when scatter is reduced to a negligible level, will be background emission shot- and flicker-noise. This holds at low analyte concentrations, i.e., near the limit of detection. Other types of noise may become dominant in extreme situations, for example in the presence of very large amounts of matrix constituents, but, as far as the power of detection of the technique is concerned, the atomiser noise will usually be the limiting one. This type of noise can be called "extrinsic", according to Alkemade [11], to distinguish it from the "intrinsic" one, which can be calculated assuming that the limiting error is that inherent in the signal, and therefore caused by the statistical fluctuations in the number of atoms probed and in the detection process. The concentrational detection limits given in the fifth column of Table 1, converted to number densities of free atoms in the atomiser volume, correspond to approximately 10^5–10^6 atoms cm^{-3}, which indicates that the technique is still very far from single atom detection capabilities.

In the case of LEI, the fluctuation (shot noise) in the natural background current, due to the number of ions and electrons present in the atomiser, appears to be the limiting noise. Calculations based upon a value of a 10 μA for the air-acetylene flame indicate that the best detection limits reported in the second column of

Table 1. Limits of detection obtained for several elements in aqueous solutions by LEI and LIF in the air-acetylene flame, by inductively coupled plasma-mass spectrometry, and by LIF and LEI with electro-thermal atomisation (ETA)

Element	LEI[a]	LIF[b]	ICP-MS[c]	LIF-ETA[d]	LEI-ETA[e]
			(ng/ml)		
Ag	0.07	4	0.04	0.0004	
As	3000		0.4		
Au	1	4	0.08		
Ba	0.2	8	0.02		
		(1.5)			
Bi	2	3	0.06		
Ca	0.03	0.8	5		
		(0.01)			
Cd	0.1	8	0.07	0.0009	
Co	0.08	200	0.01	0.003	0.25
Cr	0.2	1	0.02		0.25
Cu	0.7	1	0.03	0.003	
Fe	0.08	0.06	0.2	0.005	
Ga	0.04	0.9	0.08	0.5	
In	0.0009	0.2	0.01	0.001	0.0003[f]
Li	0.0003	0.5	0.06		
Mg	0.003	0.2	0.1		
		(0.08)			
Mn	0.02	0.4	0.04	0.004	0.05
Na	0.003	0.1	0.06	0.03	0.0003[f]
Ni	0.08	0.5	0.03	0.05	0.35
Pb	0.09	0.02	0.02	0.00003	0.03
Sb	50	50	0.02		
Sr	0.01	0.3	0.02		
		(1.5)			
Tl	0.008	0.8	0.05	0.00004	
Zn	1		0.08		

[a] Values taken from [18]
[b] Values taken from [9]. Values in parentheses taken from [17]
[c] Values taken from SCIEX commercial literature (ELAN: Elemental Analysis System), Sciex Corp., Tornhill, Ontario, Canada
[d] Values taken from [25, 27, 28–32]. These values refer to the best detection limit reported, obtained with either graphite tube or graphite cup atomisation. All values have been calculated for a sample volume of 20 μl
[e] Values taken from [22] and for a sample volume of 20 μl
[f] Values taken from [21] (average range given) and calculated for a sample volume of 20 μl

Table 1 are already close to the ultimate predictable value [16]. These limits will be degraded, however, if the sample matrix is easily ionisable, since the flame background current and its fluctuations will increase markedly.

1.2. Experimental Systems

A typical experimental set-up for fluorescence and ionisation measurements is shown in Fig. 2. In this case, an excimer laser is split in order to pump two dye lasers which counterpropagate colinearly along the flame axis in the case of ionisation measurements, or are adjusted by means of mirrors so as to provide optimum overlap in the flame or plasma for the fluorescence measurements.

Several types of lasers have been used in the literature: from cw lasers to flash lamp-pumped dye lasers, to Nd:YAG-pumped and excimer-pumped dye lasers and to copper vapour-pumped dye lasers. Nd:YAG- and excimer-pumped dye lasers are the most commonly used, due to their wavelength coverage and reliability. The best type of laser may be different according to the particular experiment at hand. For example, larger ionisation yields [see Eqs. (1 and 2)] are expected for flash lamp-pumped dye lasers (1 μs pulse duration) or cw lasers (where the duration of the interaction is dictated by the residence time of the

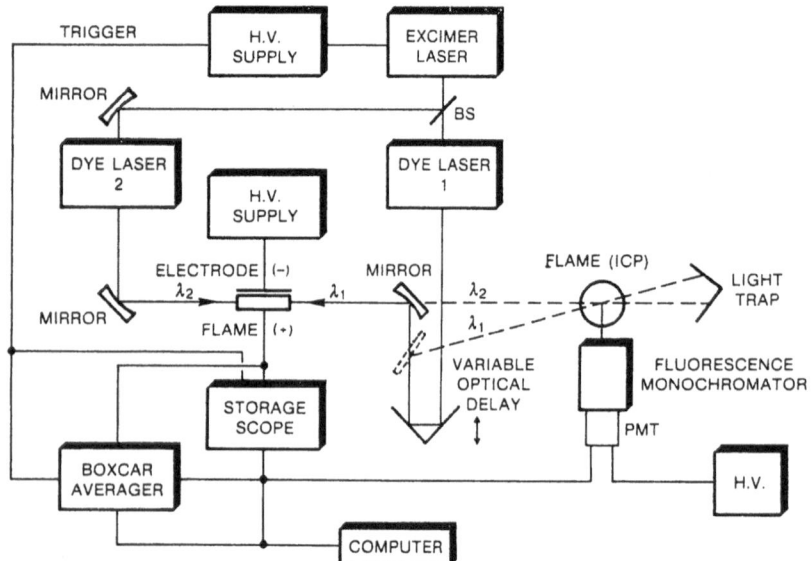

Fig. 2. Typical experimental set-up used in the fluorescence and ionisation experiments with flames and plasmas operated at atmospheric pressure. HV = High Voltage; BS = Beam Splitter; PMT = Photomultiplier Tube. (From [17], with permission)

atoms in the laser beam) but in the fluorescence approach high peak powers (associated with short pulses) are needed in order to overcome quenching collision rates which are known to be high in hydrocarbon-fuelled flames.

Flames, plasmas, and graphite furnaces have been tested as atom reservoirs. With flames, little or no modification of the conventional burner and nebulis-

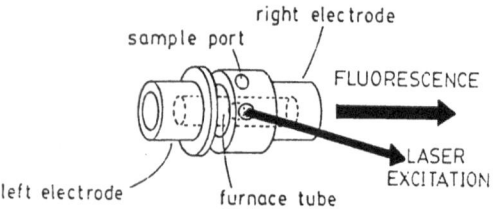

Fig. 3. Modified Perkin-Elmer HGA-500 atomiser. Two 5-mm holes were drilled through the right electrode and two 4-mm holes were drilled through the furnace tube. (From [25], with permission)

ing chamber used in the well established atomic absorption technique was necessary for either fluorescence or ionisation measurements. With graphite furnaces however, the situation is different. Here, the atomisation can be performed in a cylindrical tube or in a hollow cup. In the latter case, the fluorescence is observed just above the cup while in the former case the tube can be modified (see Fig. 3) in order to allow for the observation of the fluorescence at 90° with respect to the excitation path [24, 25]. In recent experiments [26, 27], it was shown that the fluorescence could be efficiently collected by a combination of a plane mirror, having a hole in its centre to allow exciting of the excitation beams, positioned at 45° with respect to the longitudinal axis of the tube, and a lens positioned so as to image the centre of the tube into the entrance slit of the fluorescence monochromator (see Fig. 4). With this arrangement, no modifications of the graphite tube and its inherent electrical characteristics are required.

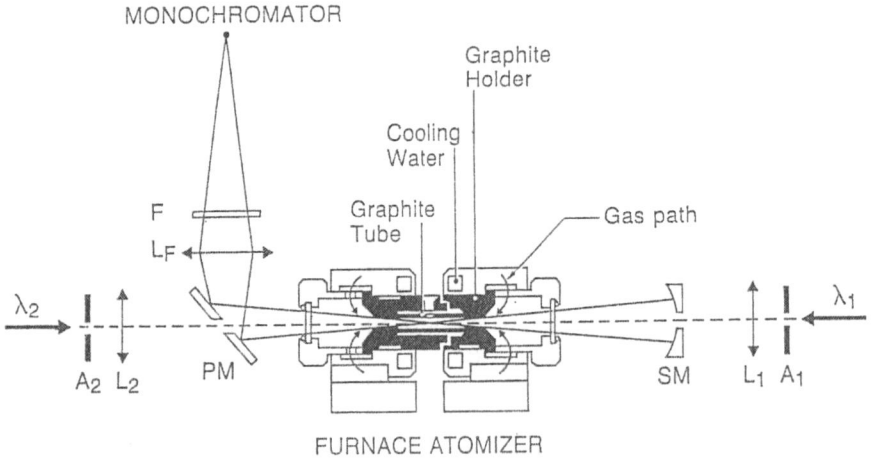

Fig. 4. Longitudinal observation of the double-resonance fluorescence excited in a graphite tube atomiser by two counter-propagating laser beams tuned at λ_1 and λ_2. A_1, A_2 = Apertures; L_1, L_2 = Collimating lenses; PM, SM = Plane Mirror, Spherical Mirror; L_F = Fluorescence Lens; F = Filter. (From [27] with permission)

In ionisation experiments with graphite cup atomisers [21], the electrode is placed above the electrically grounded graphite cup and the laser excitation set between the electrode and the cup, just as in atomic fluorescence. With graphite tube atomisation [21, 22], a wire is mounted axially and centred inside the graphite tube, the signal being measured on the wire maintained at a given potential with respect to the earthed tube wall.

The processing of the fluorescence and ionisation signals is usually done with boxcar integrator, whose gate position and width are optimised according to the duration of the laser pulse, the type of dominant noise in the system and the temporal resolution needed.

1.3. Analytical Results

1.3.1. LIF and LEI with Atmospheric Pressure Atomisers. The detection limits obtained by the technique of LIF and LEI for several elements and atomisers are collected in Table 1. These limits are referred to aqueous solutions of the pure element and not to real sample solutions, and are calculated for a signal-to-noise ratio of 3. The noise is usually calculated as the standard deviation of several "blank" measurements, obtained by running the experiment with the purest water available. If this blank gives a signal which is demonstrated to be truly due to the analyte, present as an impurity or as a contamination, the standard deviation is measured "off-line", i.e., by detuning the laser away from the resonance transition. The technique chosen as "reference" for a comparison of detection limits is that which combines the characteristic of plasma sources with mass spectrometric detection, since it indeed provides uniform, excellent sensitivity for most elements of the periodic table.

Several considerations are outlined below:

(i) LIF in the air-acetylene flame and inductively coupled argon plasma provides detection limits of the order of 1 ng/ml, with a few exceptions (in the right and in the wrong directions). This value is about 3 orders of magnitude higher than that required for applications foreseen at the beginning of this paper for which, therefore, LIF in flames does not provide the necessary sensitivity. Nevertheless, in those cases where the sensitivity is adequate, (e.g., in the direct determination of lead in blood, where the detection limit of 0.02 ng/ml given in Table 1 is amply sufficient), LIF can be successfully used and has the advantage of high spectral selectivity;

(ii) Contrary to the case of LIF, LEI in the air-acetylene flame does indeed approach detection limits in the range of pg/ml (and even below in the case of Li). As a result, for some elements, this sensitivity is adequate for the type of applications previously described. The higher sensitivity of LEI compared to LIF resides in its much higher collection efficiency and in the absence of problems due to scattering and stray light. LEI has been successfully applied to a variety of standard reference materials, especially alloy analyses;

(iii) The best detection limits provided by LIF are those obtained with electro-thermal atomisation. Most of the tabulated data refer to single-resonance fluorescence and have been calculated for 20 μl of sample solution introduced in the furnace. The absolute sensitivities are therefore in the range of femtograms. Of the twelve elements listed, lead and thallium reach detection limits of 30 and 40 fg/ml, respectively, which points to the possibility of direct detection of these elements at levels of 3 and 4 parts in 10^{12} in the original sample.

Analytical applications of LIF with electro-thermal atomisation are not very numerous. The technique has been successfully applied to the determination of Fe, Co, and Cu in soil extracts with sensitivities reaching the $10^{-9}\%$ level [28]. It is also worth mentioning that the Zeeman background correction procedure has also been adapted to LIF and graphite furnace atomisation by propagating the laser beam through the pole pieces of the electromagnet, parallel to the magnetic field [33].

(iv) Only a few elements have been tested with LEI in graphite furnaces and the results are therefore to be considered on a preliminary basis only. With properly designed furnaces and detection geometries, the sensitivity is expected to improve by several orders of magnitude.

1.3.2. LIF and RIS with Atomisation Under Vacuum. Vacuum electro-thermal atomisation in LIF has the advantage of providing a radical way to eliminate gas phase reactions and quenching, which are a cause of interference in atmospheric pressure atomisers when real samples with complex chemical composition are analysed [34]. Collisionless expansion of atoms with thermal velocities results however in a dilution of the vapour cloud with a consequent reduction in the number density of atoms present in the laser-irradiated volume and of the fluorescence signal. The best analytical advantage of the method is that the chemical composition of the matrix is irrelevant for the analytical result. Such a technique has been applied to the determination of cobalt in samples of pure tin, vegetation, and pure glass [34].

The body of literature describing RIS applications to chemical analysis with vacuum atomisation is very large and the reader is referred to [8, 12, 14, 35–37] for an exhaustive coverage of the subject. The RIS technique meets the sensitivity requirements foreseen in the introduction, since it is capable of analysing many elements present in almost any matrix at levels ranging

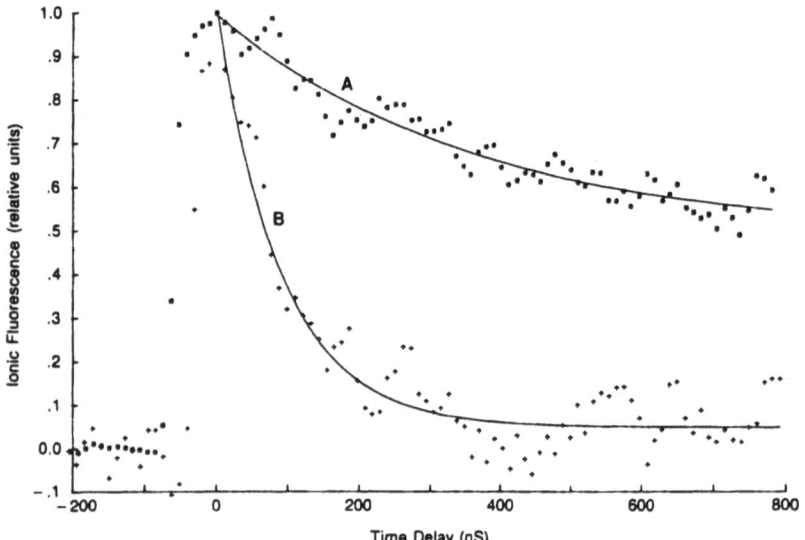

Fig. 5A, B. Effect of the presence of the argon sheath around the air-acetylene flame on the fast decay of strontium ionic fluorescence. The signals have been normalized to the same scale and the natural ionisation level has been subtracted. The net laser-produced ion level at time coincidence is 2.6 times greater with the sheath on. (A) Argon sheath on; (B) Argon sheath off. (Reproduced from [38] with permission)

from $10^{-8}\%$ (0.1 ng/g) to $10^{-11}\%$ (0.1 pg/g). In a recent application on the determination of gallium in pure solutions and in solid germanium [37], an absolute detection limit of 5 femtograms was reached. This limit, converted into the units given in Table 1, corresponds to a concentration of 0.0003 ng/ml.

1.4. Diagnostic Applications of LIF and LEI

This section would cover an immense amount of material if all the separate diagnostic applications of LIF and LEI were reported.

To mention just a few, LIF offers the possibility of measuring the quantum efficiency of a particular fluorescence transition, the excitation temperature of flames and plasmas, the lifetime of the excited levels, the transition probability and the total number density of the emitting atoms under saturated conditions.

Likewise, LEI can provide flame rise velocity data, electric field distribution, mobilities and diffusion processes in flames, as well as lifetimes of excited levels. Since the coverage of these applications is outside the scope of the present paper, it seems worthwhile to discuss below only a few examples in which the *simultaneous* use of the fluorescence and ionisation techniques in flames and plasmas has proven to be a very informative diagnostic tool.

1.4.1. Ion Decay Kinetics.

The fate of the laser-created ions in the atomiser can be followed by the laser induced ionic fluorescence technique. This optical detection of LEI was applied to the study of the decay of strontium ions in the air acetylene flame [38] and in the inductively coupled argon plasma [39].

In the experiment, two excimer-pumped dye lasers were used. The strontium atoms were photoionised by tuning one dye laser in resonance with an atomic transitions and using the excimer as the photoionising laser. The resulting ions were then interrogated with a third laser beam, tuned to a strontium ionic transition and delayed by variable amounts from the first two lasers. In this way, the influence of the flame composition on the disappearance of the ions could be studied as well as the effects of easily ionisable elements (Li, Cs, K) on the recombination rate between the electrons and the ions. It was found that 85% of the laser-produced strontium ions decay with an exponential time constant of 58 ns while the remaining 15% decay at a much slower rate. The former decay was attributed to flame chemistry and the last one to ion-electron recombination.

As an example of the results obtained, Fig. 5 shows the striking difference in the decay rate of the ions when an argon sheath is present around the flame. The sheath considerably decreases the concentration of OH radicals at the flame border, where the lasers were purposely located.

1.4.2. Fluorescence Dip Spectroscopy.

As already pointed out in Sect. 1.1.6, if the resonance fluorescence signal resulting from the first excitation step in a double-resonance fluorescence or ionisation experiment is monitored while spectrally scanning the second excitation step, a dip will be observed due to the depletion of the population of the first excited level. In a simultaneous fluorescence-ionisation experiment, when the first excitation transition is fixed and the second scanned in wavelength, a laser-enhanced ionisation spectrum will be recorded together with a fluorescence dip spectrum [40]. As an example, the ionisation enhancement and the corresponding fluorescence dip for the lithium atoms in the air-acetylene flame are shown in Fig. 6. It can be shown that the information content of the fluorescence dip is similar to that of the saturated fluorescence signal. The dip in fact will increase by increasing the laser spectral energy density of the second step and will reach a plateau at high densities. From this saturation curve, the saturation parameter and the quantum efficiency relative to the second fluoresence transition can be evaluated. A particularly attractive feature of the time-resolved fluorescence dip is that it would allow the direct evaluation of the absorption cross section of the second excitation step, or, in the case of photoionisation, the photoionisation cross section.

2. Thermal Lensing Spectrophotometry (TLS)

This technique belongs to the field of *thermo-optical* techniques, which have significantly improved absorbance measurements in comparison with standard transmission methods. This was possible because of the peculiar characteristics of the laser radiation, in particular its high spatial coherence. In thermo-optical methods, the index of refraction of an absorbing medium changes as a result of the temperature rise caused by the absorption of the laser radiation. The heated sample acts therefore as an optical element.

As stated in a detailed review [42], the lexicon of thermo-optical techniques is surprisingly rich, since a classification is often made by analogy with the conventional optical elements, e.g., a lens in thermal lens calorimetry, a grating in *thermal diffraction*, a prism in *thermal deflection* and so on.

The theoretical aspects of thermo-optical methods are in general quite complicated, involving a thorough description of the propagation of the laser beam, the production of a profile of the refractive index in the sample and the mechanism of dissipation of heat in the medium. Such description is beyond the scope of the present paper and the reader is referred to some pertinent literature [42–49]. On the other hand, it was felt appropriate to discuss some applications of the TLS method, for example, the detection of trace metals in solutions and the study of the chemical equilibria occurring at low concentrations of uranyl ions in water.

2.1. Determination of Trace Metals in Solution

The sensitivity of the TLS technique is better judged on the basis of the minimum observable absorbance rather than on the minimum detectable concentration. For a 1 cm absorption path length, a detection limit of 10^{-7} cm^{-1} can be reasonably considered to be within the reach of the technique. The molar absorptivity of a strongly absorbing, non fluorescent complex can be as

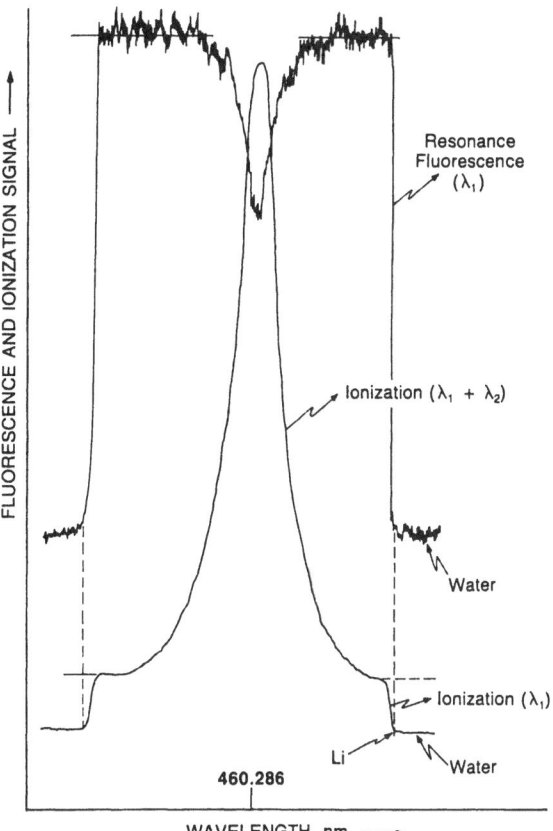

Fig. 6. Simultaneous recording of the fluorescence and ionisation signals for both single-step and two-step excitation. The wavelength of the second laser, which is spectrally scanned, is indicated in the abscissa. The background fluorescence and ionisation level due to water are indicated. When λ_2 is tuned into resonance with the second step, an enhancement in the ionisation signal and a dip in the fluorescence signal are observed. (Reproduced from [41], with permission)

high as 10^4 l mol^{-1} cm^{-1}, which then corresponds to a concentrational detection limit of 10^{-11} M. For an atomic weight of 100, this corresponds to 1 pg/ml, which compares favourably with the *best* figures quoted in Table 1.

Both metals and non metals have been determined by TLS [42, 49]. Solvent extraction was applied in all cases. In a recent experiment, the technique was applied to several elements complexed with dithizone and extracted in carbon tetrachloride [50]. All ions investigated (Cu^{2+}, Zn^{2+}, Ag$^+$, Cd^{2+}, Hg^{2+}, Pb^{2+}, and Bi^{3+}) gave detection limits similar to those reported in Table 1 for the technique of inductively coupled plasma-mass spectrometry. For Cd^{2+}, the concentrational detection limit was 8 pg/ml, while the mass limit in the laser probe volume was 4 fg.

Solvent background absorbance will be a major limiting factor in the technique. In this regard, TLS can be used in the *differential* mode [51], by taking

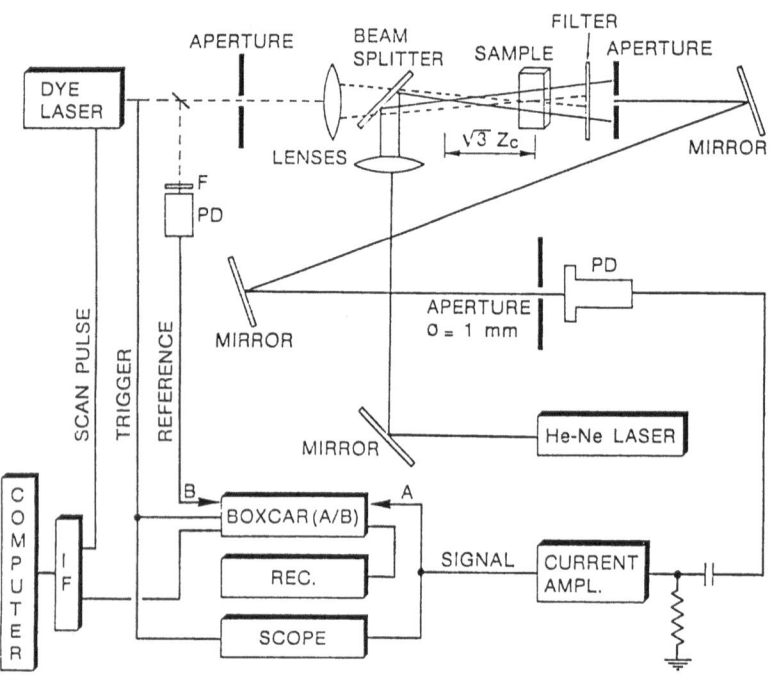

Fig. 7. Dual-beam experimental set-up for Thermal Lensing Spectrophotometry with pulsed lasers. F: Filter; PD: Photodiode; IF: Interface. (Reproduced from [53], with permission)

advantage of the fact that the thermal lens collimates or defocuses the beam according to its position with respect to the beam waist; the effect can therefore be cancelled when *two* thermal lenses are placed symmetrically about the beam waist. This differential set-up was used for the study of lanthanides and actinides in solution for both cw and pulsed lasers [52, 53].

2.2. Studies of Speciation and Chemical Equilibria

These studies are of particular relevance when applied to traces of actinide ions in water solutions. In fact, a knowledge of the migration mechanism of these ions in ground waters (which act as an efficient carrier for their transport) is essential to elucidate radionuclide reactions in geochemical processes such as leaching from nuclear waste repositories. There are, however, only very few sensitive analytical methods that can be used for their direct determination and speciation. TLS and photoacoustic spectroscopy [54] (which is also capable of measuring small absorbances) were recently applied for that purpose [53, 55–58].

In our laboratory, we have developed a dual-beam system, based on pulsed laser excitation, which is used for the study of the complexation behaviour of uranium in a carbonate-perchlorate system. The experimental set-up is shown in Fig. 7. The dye laser beam and the probe beam can be independently focussed by means of two lenses, the absorbing cuvette being placed at the waist of the excitation beam. The waist of the probe beam is adjusted to be at the position of maximum sensitivity, i.e. at $(3)^{1/2}Z_c$ from the cell,

where Z_c is the confocal distance of the probe beam. The intensity at the probe beam centre is measured with a photodiode located behind a pinhole of 1.5 mm diameter. The photodiode current, converted into voltage and amplified, is fed into a digital storage oscilloscope and into a boxcar integrator, interfaced with a computer.

The sensitivity of the system was checked with a 5×10^{-7} M solution of U(VI), of molar absorptivity equal to 25 l mol^{-1} cm^{-1}. The detection limit, for a signal-to-noise ratio of 3, was found to be 7.5×10^{-7} cm^{-1}. This sensitivity allowed the study of the absorption spectrum of the ion, characteristic of its degree of oxidation and chemical form, at concentrations $(4 \times 10^{-6}$ M) which are totally inaccessible with ordinary spectrophotometry [53, 56].

The same system has been recently used to study the mononuclear hydrolysis of 10^{-5} M concentrations of U(IV). The TLS measurements were made by using a reactor vessel which was connected via glass tubing and a piston burette, acting as a pump, to the absorbing cuvette, which was then kept fixed in place during the overall procedure. The fit of the experimental results to the theoretical model was satisfactory.

3. Concluding Remarks

Fluorescence and ionisation techniques in atmospheric pressure atomisers are indeed capable of approaching the extreme sensitivity needed for the detection of several trace metals in a variety of materials.

However, there is still need for improved detection powers, especially if one takes into account the fact that the detection limits reported are calculated on statistical grounds and that the *actual* analysis must be performed at a concentration which is at least 10 times higher than the quoted limit of detection.

For both LIF and LEI, work must be directed towards the application of new atomisers which could replace flames and plasmas, even at the expense of the simplicity of operation and speed of analysis which are typical of these systems. Electro-thermal atomisation at atmospheric pressure provides the highest sensitivity, while vacuum atomisation is the best approach in terms of freedom from matrix effects and therefore of real analytical usefulness.

Other atomisation techniques will be more fully characterized in analytical practice, e.g., laser ablation of sample targets and glow-discharge sputtering devices operated at low pressure in a pulsed mode.

Thermal lensing spectrophotometry and photoacoustic spectroscopy are good examples of how lasers are used to perform experiments which are impossible with conventional analytical methods. In this respect, they provide one of the best applications of lasers to real analytical problems.

In conclusion, laser-based methods of chemical analysis are not expected to replace conventional methods, whenever the sensitivity of these methods is sufficient and/or a given analytical procedure well established. It must be the necessity to solve difficult analytical problems which forces the analyst to contemplate the use of laser techniques, irrespective of how complex, costly or time-consuming they are.

Fundamental and applied research in analytical laser spectroscopy will therefore continue.

References

1. G.M. Hieftje: J. Chem. Educ. **59**, 900 (1982)
2. J.D. Winefordner, M.S. Epstein: In *Physical Methods of Chemistry, Vol. IIIA: Determination of Chemical Composition and Molecular Structure* ed. by B.W. Rossiter, J.F. Hamilton (Wiley, New York 1987)
3. J.A.C. Broekaert, G. Tölg: In *Inductively Coupled Plasma Emission Spectroscopy, Part II: Applications and Fundamentals* ed. by P.W.J.M. Boumans (Wiley, New York 1987)
4. N. Omenetto (ed.): *Analytical Laser Spectroscopy* (Wiley, New York 1979)
5. G.M. Hieftje, J.C. Travis, F.E. Lytle (eds.): *Lasers in Chemical Analysis* (Humana Press, Clifton, NJ 1981)
6. D.S. Kliger (ed.): *Ultrasensitive Laser Spectroscopy* (Academic, New York 1983)
7. J.J. Snyder, R.A. Keller (eds.): *Ultrasensitive Laser Spectroscopy*. J. Opt. Soc. Am. B **2**, 1427 (1985)
8. V.S. Letokhov (ed.): *Laser Analytical Spectrochemistry* (Adam Hilger, Bristol 1985)
9. E.H. Piepmeier (ed.): *Analytical Applications of Lasers* (Wiley, New York 1986)
10. W. Demtröder: *Laser Spectroscopy* (Springer, Berlin, Heidelberg 1982)
11. C.Th.J. Alkemade: In *Analytical Applications of Lasers*, ed. by E.H. Piepmeier (Wiley, New York 1986) Chap. 4
12. V.I. Balykin, G.I. Bekov, V.S. Letokhov, V.I. Mishin: Sov. Phys. Usp. **23**, 651 (1980)
13. G.S. Hurst, M.G. Payne, S.D. Kramer, J.P. Young: Rev. Mod. Phys. **51**, 767 (1979)
14. V.S. Letokhov: *Laser Photoionisation Spectroscopy* (Academic, Orlando 1987)
15. P. Camus (ed.): *Optogalvanic Spectroscopy and its Applications*. J. Phys. **44**, Coll. C7, Suppl. 11 (1983)
16. J.C. Travis, G.C. Turk, J.R. DeVoe, P.K. Schenck, C.A. Van Dijk: Progr. Anal. Atom. Spectr. **7**, 199 (1984)
17. N. Omenetto, B.W. Smith, L.P. Hart: Fresenius Z. Anal. Chem. **324**, 683 (1986)
18. G.C. Turk: J. Anal. At. Spectr. **2**, 573 (1987)
19. F.M. Curran, K.C. Lin, G.E. Leroi, P.M. Hunt, S.R. Crouch: Anal. Chem. **55**, 2382 (1983)
20. O. Axner, I. Magnusson, J. Petersson, S. Sjöström: Appl. Spectrosc. **41**, 19 (1987)
21. I.V. Bykov, A.B. Skvortsov, Yu. G. Tatsii, N.V. Chekalin: In [15] p. 345
22. I. Magnusson, S. Sjöström. M. Lejon, H. Rubinsztein-Dunlop: Spectrochim. Acta **42B**, 713 (1987)
23. G.C. Turk, R.L. Watters Jr.: Anal. Chem. **57**, 1979 (1985)
24. K. Dittrich, H.J. Stark: J. Anal. At. Spectr. **1**, 297 (1986)
25. J.P. Dougherty, F.R. Preli, R.G. Michel: J. Anal. At. Spectr. **2**, 429 (1987)
26. D. Goforth, J.D. Winefordner: Talanta **34**, 290 (1987)
27. N. Omenetto, P. Cavalli, M. Broglia, P. Qi, G. Rossi: J. Anal. At. Spectr. **3**, 231 (1988)
28. M.A. Bolshov, A.V. Zybin, I.I. Smirenkina: Spectrochim. Acta **36B**, 1143 (1981)
29. H. Falk, J. Tilch: J. Anal. At. Spectr. **2**, 527 (1987)
30. F.R. Preli Jr., J.P. Dougherty, R.G. Michel: Anal. Chem. **59**, 1784 (1987)
31. K. Dittrich, J. Stark: J. Anal. At. Spectr. **2**, 63 (1987)
32. J. Vera, M. Leong, B.W. Smith, N. Omenetto, J.D. Winefordner: In preparation
33. J.P. Dougherty, F.R. Preli Jr., J.T. McCaffrey, M.S. Seltzer, R.G. Michel: Anal. Chem. **59**, 1112 (1987)
34. M.A. Bolshov, A.V. Zybin, V.G. Koloshnikov, I.A. Mayorov, I.I. Smirenkina: Spectrochim. Acta **41B**, 487 (1986)
35. G.R. Bekov, V.S. Letokhov: Appl. Phys. B **30**, 141 (1983)
36. G.I. Bekov, V.S. Letokhov, V.N. Radaev: J. Opt. Soc. Am. B **2**, 1554 (1985)
37. G.I. Bekov, V. Radaev, J. Likonen, R. Zilliacus, I. Auterinen, E. Lakomaa: Anal. Chem. **59**, 2472 (1987)
38. G.C. Turk, N. Omenetto: Appl. Spectrosc. **40**, 1085 (1986)
39. G.C. Turk, O. Axner, N. Omenetto: Spectrochim. Acta **42B**, 873 (1987)
40. N. Omenetto, G.C. Turk, M. Rutledge, J.D. Winefordner: Spectrochim. Acta **42B**, 807 (1987)
41. B.W. Smith, L.P. Hart, N. Omenetto: Anal. Chem. **58**, 2151 (1986)
42. N.J. Dovichi: CRC Crit. Rev. Anal. Chem. **17**, 357 (1987)
43. C. Hu, J.R. Whinnery: Appl. Opt. **12**, 72 (1973)
44. A.J. Twarowski, D.S. Kliger: Chem. Phys. **20**, 253 (1977)
45. R.T. Bailey, F.R. Cruickshank, D. Pugh, W. Johnstone: J. Chem. Soc., Faraday Trans. **2**, 76, 633 (1980)
46. H.L. Fang, R.L. Swofford: In *Ultrasensitive Laser Spectroscopy*, ed. by D.S. Kliger (Academic, New York 1983)

47. S.J. Sheldon, L.V. Knight, J.M. Thorne: Appl. Opt. **21**, 1663 (1982)
48. S.E. Bialkowski: Appl. Opt. **23**, 2792 (1973)
49. J.M. Harris: In *Analytical Applications of Lasers*, ed. by E.H. Piepmeier (Wiley, New York 1986) Chap. 13
50. G. Ramis Ramos, M.C. Garcia Alvarez-Coque, B.W. Smith, N. Omenetto, J.D. Winefordner: Appl. Spectrosc. **42**, 341 (1988)
51. N.J. Dovichi, J.M. Harris: Anal. Chem. **52**, 2338 (1980)
52. T. Berthoud, N. Delorme: Appl. Spectrosc. **41**, 15 (1987)
53. N. Omenetto, P. Cavalli, G. Rossi, G. Bidoglio, G.C. Turk: J. Anal. At. Spectrosc. **2**, 579 (1987)
54. D. Betteridge, P.J. Meylor: CRC Crit. Rev. Anal. Chem. **14**, 267 (1984)
55. R. Stumpe, J.I. Kim, W. Schrepp, H. Walther: Appl. Phys. B **34**, 203 (1984)
56. G. Bidoglio, G. Tanet, P. Cavalli, N. Omenetto: Inorg. Chim. Acta **140**, 293 (1987)
57. P.M. Pollard, M. Liezers, J.W. Edwards, J.W. McMillan: XXV Colloquium Spectroscopicum Internationale, Toronto, Canada (1987), Book of Abstracts, p. 118
58. C. Moulin, T. Berthoud, N. Delorme, P. Mauchien: Radiochim. Acta (1988) (in press)

Appl. Phys. B 46, 221–236 (1988)

Applied Physics B

Photo-
physics
and Laser
Chemistry

© Springer-Verlag 1988

Laser Spectroscopy for Studying Chemical Processes

J. Wolfrum

Physikalisch-Chemisches Institut, Universität, Im Neuenheimer Feld 253,
D-6900 Heidelberg, Fed. Rep. Germany

Received 11 March/Accepted 15 March 1988

Abstract. In recent years, various methods have been developed to observe and to influence the course of chemical reactions using laser radiation. By selectively increasing the translational, rotational, and vibrational energies and by controlling the relative orientation of the reaction partners with tunable infrared and UV lasers, direct insight can be gained into the molecular course of the breaking and re-forming of chemical bonds. As examples for the application of lasers in chemical synthesis the production of monomers and catalysts is discussed. The application of linear and nonlinear laser spectroscopic methods, such as laser-induced fluorescence (LIF), Coherent anti-Stokes Raman Scattering (CARS), infrared-absorption measurements with tunable diode and molecular lasers is described for non-intrusive observation of the interaction of transport processes with chemical reactions used in industrial processes with high temporal, spectral and spatial resolution. Finally the application of a UV laser microbeam apparatus in genetic engineering for laser-induced cell fusion, genetic transformation of plant cells as well as diagnosis of human diseases by laser-microdissection of chromosomes is described.

PACS: 82.20, 82.30, 82.50, 82.80, 87

Irradiation with light can have a considerable influence on the course of chemical reactions. The best known example is the photosynthesis in plants by sunlight. In addition, electromagnetic radiation is the most important aid to determining structure, properties, and behaviour of chemically reacting substances. In spite of great successes, spectroscopy with conventional light sources has not been able to answer many interesting questions. As in other areas of science and technology, the laser has also created numerous new possibilities for the investigation of chemical processes. Particularly the introduction of tunable laser light sources, such as the dye laser [1] and the development of linear and nonlinear optical techniques such as laser-induced fluorescence (LIF), resonant multiphoton ionization (REMPI), coherent anti-Stokes Raman spectroscopy (CARS), laser Raman spectroscopy on surfaces (SERS), photoacoustic spectroscopy (PAS), laser magnetic resonance spectroscopy (LMR) and Doppler-free absorption spectroscopy, allow virtually every spectroscopic state of an atom or molecule to be observed with high resolution, from the far infrared with wavelengths of several millimetres to wavelengths in the nanometre range in the vacuum ultraviolet [2]. The high sensitivity makes observation of single atoms and molecules possible. Furthermore, laser spectroscopy can provide non-intrusive observation of rapidly changing chemical reactions, such as combustion processes, with high temporal, spectral, and spatial resolution. In this contribution the application of laser techniques to observe and to stimulate chemical reactions will be described in four different areas: chemical kinetics, chemical synthesis, industrial chemical reactors and genetic engineering.

1. Microscopic Dynamics of Elementary Chemical Reactions

The strong dependence of the rate of chemical reactions on the energy supplied is one of the most important observations for the chemist. Often, the

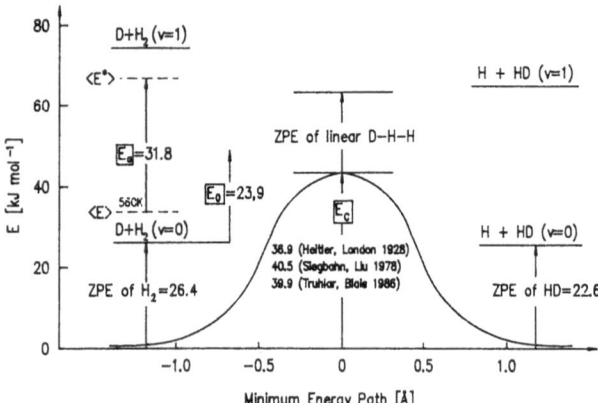

Fig. 1. Characteristic energy requirements of the hydrogen exchange reaction. E_a is the Arrhenius activation energy, E_0 the minimum collision energy that leads to reaction (threshold energy), and E_c the height of the potential energy barrier of the reaction in the vibrational ground state, $\langle E \rangle$ is the average thermal energy of all binary collisions per unit time and $\langle E^* \rangle$ the average kinetic energy of all collisions per unit time that lead to reaction. ZPE is the zeropoint energy of vibration

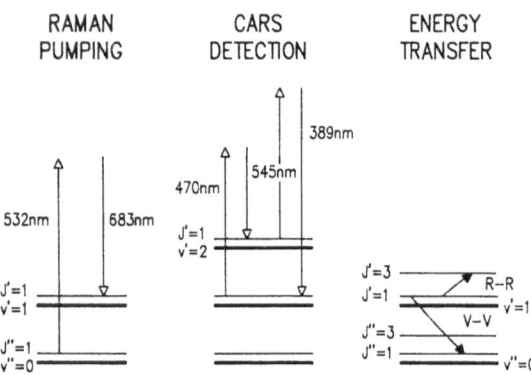

Fig. 2. Energy level diagram for the excitation and detection of state-selected hydrogen molecules by Raman and CARS spectroscopy

energy of the reaction partners can be represented by one temperature and the dependence of the reaction rate on temperature is described by the now nearly one-hundred-year-old Arrhenius equation [3]. The Arrhenius parameters obtained in this way do not, however, provide any information on the respective contribution of the degree of freedom of the participating reaction partners to overcoming the energy barrier of the reaction. In the following examples, reactions in the gas phase are used to show how information on microscopic details of breaking and re-forming of bonds in chemical reactions can be obtained in experiments with lasers. Such information can be used to influence chemical reactions much more selectively than by classical thermal heating of the reactants.

The reaction of hydrogen molecules with hydrogen atoms or their isotopes, as the simplest example of a

bimolecular reaction of neutral particles, is particularly suitable for a theoretical investigation of the influence of selective excitation of reacting species. The energy of one vibrational quantum of the hydrogen molecule considerably exceeds the Arrhenius activation energy (E_a), the threshold energy (E_0) as well as the height of the potential energy barrier (E_c) of the reaction in the vibrational ground state (see Fig. 1). E_c was first calculated quantum mechanically by London [4] more than half a century ago. Classical methods for the study of reaction kinetics are difficult to apply to this reaction, because known concentrations of vibrationally excited hydrogen molecules have to be produced and detected. Due to the lack of a dipole moment and an electronic absorption spectrum in the vacuum ultraviolet, state-selective studies using spectroscopic methods were difficult to perform for a long time, before laser methods became available. Figure 2 shows the excitation and detection scheme of a laser experiment for energy transfer and state-selective reaction studies of hydrogen molecules. Stimulated Raman pumping is employed to populate H_2 ($v'' = 1$, $J'' = 1$) selectively in the electronic ground state of hydrogen within a 10 ns laser pulse. The time-dependent populations in rotational and vibrational levels in hydrogen and isotopic modifications can be probed by coherent anti-Stokes Raman spectroscopy (CARS). In the experimental arrangement [5] 50% of the energy output of a linearly polarized frequency-doubled Nd:YAG laser (Quanta Ray DCR1A, at 532 nm) is focussed into a Raman cell containing a hydrogen-helium mixture with partial pressures of 20 bar and 10 bar, respectively. The helium is used to reduce the pressure-dependent line shift of the Stokes line. Stimulated Stokes Raman radiation is generated in forward and backward directions. Due to the phase conjugation effect in stimulated Raman scattering, the backward beam displayed a more homogeneous intensity distribution over the beam cross section and a smaller divergence than the forward scattered beam. Both beams are focussed into the centre of the reaction cell with a beam waist of about 200 μm diameter for both fundamental and Stokes beams. The rate constants obtained from this experiment for the relaxation processes

$$H_2(v=1, J=1) \underset{k_{31}}{\overset{k_{13}}{\rightleftarrows}} H_2(v=1, J=3)$$

of $k_{13} = 2.2 \times 10^{-12}$ cm^3/s and $k_{31} = 1.4 \times 10^{-11}$ cm^2/s are in good agreement with measurements using LIF spectroscopy in the VUV spectral region for H_2 (v, J) detection [6]. In a similar way, the vibrational energy transfer from $H_2(v=1)$ molecules can be studied. The diffusion of excited H_2 and HD out of the CARS beam strongly influences the time evolution of the CARS signal. To estimate the influence of diffusion, an analytical expression for the solution of the kinetic equations coupled with transport processes is required. From such modelling calculations the rate constants for the vibrational energy exchange processes $H_2(v=1) + HD(v=0) \rightleftarrows H_2(v=0) + HD(v=1) + \Delta E = 469.4\,cm^{-1}$ of 1.9×10^{-13} cm^3/s in the exothermal and 1.4×10^{-14} cm^3/s in the endothermal direction are obtained.

The CARS detection system provides an ideal method for monitoring directly reactants and products in the $D + H_2(v=1)$ reaction. The reaction is followed in a discharge flow system, where the atoms and $H_2(v=1)$ molecules were generated by microwave discharges [7]. $HD(v=1)$ and $HD(v=0)$ molecules can be formed in adiabatic and non-adiabatic reaction pathways in this reaction. Information on the competition of reactive and inelastic channels can be obtained by monitoring the decrease of $H_2(v=1)$ in the presence of D atoms corrected for the energy transfer process described above. The experimental results obtained so far indicate about equal importance of inelastic and reactive channels as well as a predominance of the adiabatic over the non-adiabatic reactive pathways (see Fig. 3). The experiments show that when the H_2 molecule is excited to the first vibrational state, only about a third of the vibrational energy is used to overcome the potential energy barrier E_c [8, 9]. Thus the reaction of vibrationally excited hydrogen molecules still shows an energy barrier, whose height can be predicted both from classical and from quantum mechanical calculations of the course of reaction on the "ab initio" potential surface, in good agreement with experiments [7, 8]. As shown in the Arrhenius diagram of Fig. 3, only at low temperatures can significant differences between classical and quantum predictions of the reaction rate be expected. Such experiments are underway at the moment in our laboratory.

The results obtained from the study of the hydrogen exchange reaction on the efficiency of a selective vibrational excitation cannot be simply transferred to similar reactions. Thus investigations of the similarly thermoneutral reaction $^{37}Cl + H^{35}Cl(v=1) \rightarrow H^{37}Cl(v=1,0) + {}^{35}Cl$ demonstrate the very effective use of the vibrational energy for overcoming the energy barrier [10]. The high efficiency with which the vibrational energy is used to overcome the energy barrier causes the chemical reaction, with increasing temperature, to proceed preferentially via the excited vibrational states. Thus, the temperature dependence of the reaction deviates significantly from that predicted by an Arrhenius equation [11]. The dynamics of the reaction has been simulated by classical, semiclassical, and quantum-mechanical calculations [10]. Par-

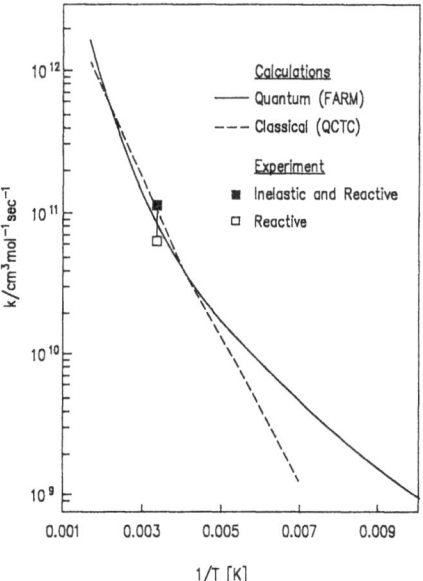

Fig. 3. Arrhenius diagram for the temperature dependence of the rate for the reaction $D + H_2 (v=1) \rightarrow HD (v=1,0) + H$ [8]

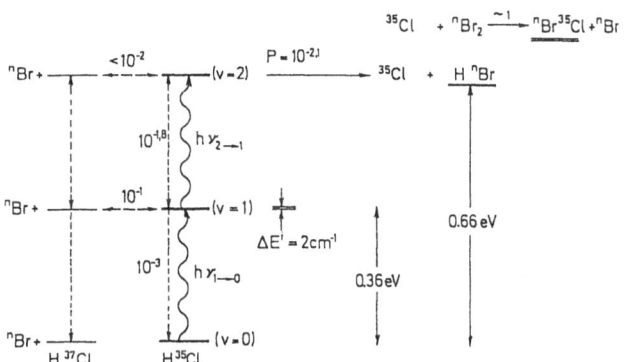

Fig. 4. Energy diagram for the reaction of Br atoms with isotope-selectively excited HCl molecules. P gives the probability for the transfer of vibrational energy between laser-excited $H^{35}Cl$ and $H^{37}Cl$ and for the course of a reaction in the various collision processes, respectively. The Br atoms react preferentially with $H^{35}Cl$ ($v=2$). The resulting ^{35}Cl atoms further react with molecular bromine to give easily separated $Br^{35}Cl$. ΔE^i is the energy difference between $H^{35}Cl$ ($v=1$) and $H^{37}Cl$ ($v=1$)

ticularly in very strongly endothermic reactions, vibrational excitation can be used efficiently for increasing the rate of the reaction. Thus, in the reaction [12] $Br + HCl(v) \rightarrow HBr(v=0) + Cl$ an increase in the reaction rate constant by more than 10 orders of magnitude can be obtained by vibrationally exciting the HCl molecule from $v=0$ to $v=2$. This dramatic increase in the rate constant can be used to separate chlorine isotopes (Fig. 4). ^{35}Cl atoms from HCl can be transformed in this way into $Br^{35}Cl$, which, because of its different chemical and physical properties, can be readily separated from the reaction mixture.

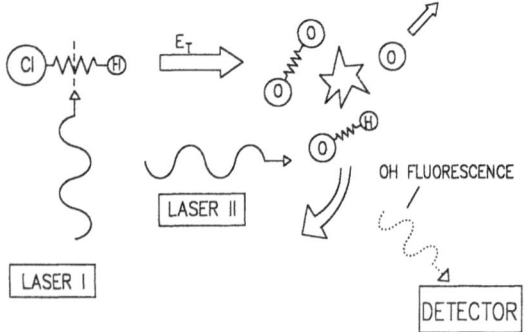

Fig. 5. Schematic for the investigation of reactions of translationally hot hydrogen atoms

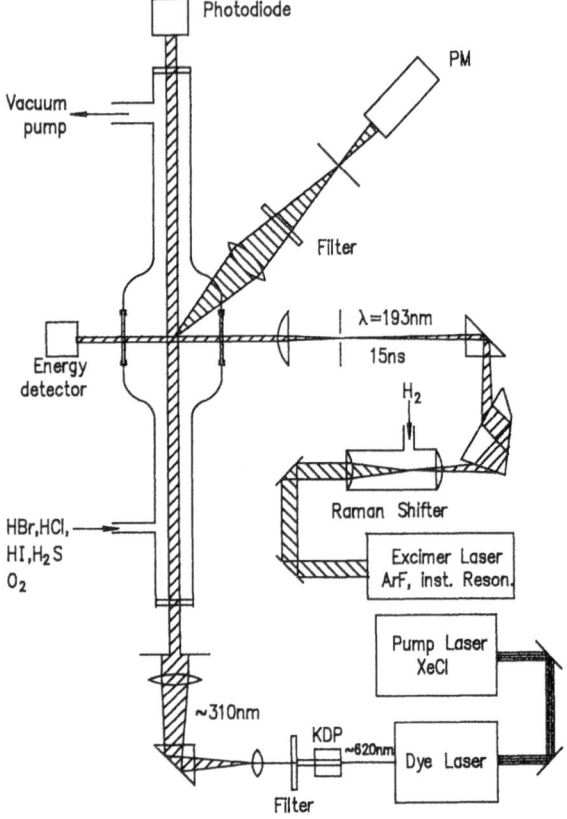

Fig. 6. Experimental arrangement for the study of reactions with translationally hot atoms and radicals by combination of excimer laser photolysis and LIF product detection

Despite the large number of elementary reactions taking place in the oxidation of hydrocarbons, important parameters of the combustion process are controlled by few elementary reactions. Sensitivity analysis shows that calculated flame velocities are relatively independent of reactions specifically for the oxidation of the fuel molecules. However, there is a strong influence on the calculated flame velocity from unspecific reactions, such as the reaction of hydrogen atoms

with oxygen molecules $H + O_2 \rightarrow OH + O$ [13]. This endothermic reaction leads to the formation of two radicals and is therefore a very important chain-branching step. As shown in Fig. 5, the dynamics of such an elementary reaction with a high energy barrier can be studied in microscopic detail by combining translationally hot atom formation from laser photolysis with time-, state- and orientation-resolved product detection by laser-induced fluorescence spectroscopy. The apparatus is depicted in Fig. 6. Two laser beams are directed perpendicular through a flow reactor equipped with long side arms to reduce the scattered light from the dye laser analysis pulse. Fluorescence light is detected as a function of the dye laser wavelength through emission optics and a filter by a photomultiplier. The experiments show [14] that the major part of the relative translational energy of the reactants is converted into rotational energy of the product OH, in agreement with the results of quasi-classical trajectory calculations [15].

The observed rotational energy distributions give interesting microscopic details on the molecular dynamics of these elementary steps. Spin-orbit and orbital-rotation interactions in the OH radical cause fine structure splittings for each rotational level. Each of these fine structure levels can be probed by different rotational sub-bands. The two OH spin states $^2\Pi_{1/2}$ and $^2\Pi_{3/2}$ are, within experimental error, equally populated. The Λ-doublet fine structure states show a clear preference for the lower energy $\Pi^+(A')$ component. The experimental results indicate that break-up of the reaction complex generates forces in a plane containing the bond to be broken. The OH radical rotates in that plane and J_{OH} is perpendicular to it and to the broken bond. This picture is consistent with a preferential planar exit channel in these reactions. This could also be directly demonstrated by using polarized photolysis and analysis laser beams [16]. The physical difference between the two Λ-doublet components $\Pi^+(A')$ and $\Pi^-(A'')$ arises from interaction of the electronic spin-orbit momentum with the rotation of the molecule. For fast rotation of the OH radical, the unpaired electron in the p orbital of the oxygen is no longer able to follow the movement of the atomic nuclei. If the p orbital lies in the OH rotational plane, the electron distribution on the oxygen atom changes, becoming increasingly spherical. In contrast, for a $\Pi^-(A'')$ configuration, the oxygen atom moves in the nodal plane of the p orbital and thus continues to "see" a dumb-bell-shaped electron environment, even for fast rotation. This leads to a splitting of the energies of the $\Pi^+(A')$ and $\Pi^-(A'')$ configurations, which selectively increases with increasing rotational energy. At 1.8 eV collision energy, about three OH radicals were found in the $\Pi^+(A')$ state for each OH radical in the

$\Pi^-(A'')$ state. This shows that the unpaired electron formed after bond cleavage of O_2 stays in an orbital in the rotational plane of the OH radical. During the reaction, most of the HO_2 complexes do not rotate out of the initial plane (see Fig. 7), which can be understood on kinematic grounds [17]. Experimentally, a total reaction cross-section of 0.42 ± 0.2 Å^2 at $E = 2.6$ eV is found [14]. The theoretical reactive cross-section obtained under these conditions by quasi-classical trajectory calculations [15, 18] on the Melius and Blint [19] surface is 0.38 Å^2. These numbers cannot be compared directly, because the multiplicity of the $^2A''$ surface and the influence of the first electronic excited state of HO_2 are not taken into account. The observed discrepancies may be attributed to a reduction of the calculated reaction cross-section due to a "rigid" character and a barrier of 8 kJ mol^{-1} in the Melius-Blint surface for dissociation of the HO_2 in the reaction $HO_2 + M \rightarrow H + O_2 + M$ [20]. Later calculations [21, 22] reduce this barrier. Also for the reaction (-1) $O + OH \rightarrow O_2 + H$ the Melius-Blint surface apparently overestimates the long-range O–OH attraction, while the Quack-Troe interpolation scheme [20] leads to better agreement with the experimental values at low temperature if the two lowest electronic states of the HO_2 radical are taken into account. Calculated rate coefficients obtained by using this theoretical cross sections from the surface [18] are in agreement with shock tube measurements for k_1 by Schott [23]. However, recent shock tube experiments [24, 25] using time-resolved atomic resonance line absorption give higher values for k_1, in agreement with the reactive cross-sections obtained in state-selective experiments. This example shows that even for a very simple chemical elementary step in combustion more work has to be done on the potential energy surface to obtain satisfactory agreement between the results from quantum chemistry and state-selective and thermal experiments using laser stimulation and detection.

2. Chemical Synthesis with Lasers

It would seem obvious to use the methods described for the selective control of chemical reactions by vibrational excitation for making new products or improving synthetic procedures. Under the typical conditions of industrial chemical processes (high pressures, reactions in the liquid phase, large molecules), a selected molecular state cannot, in most cases, be held stable until the reaction stage. The exchange of energy between the degrees of freedom within a molecule and between various molecules usually occurs on the 10^{-14} to 10^{-10} s time-scale and is therefore generally considerably faster than the reaction itself. On the other hand, nonthermal isomer and isotope distributions as

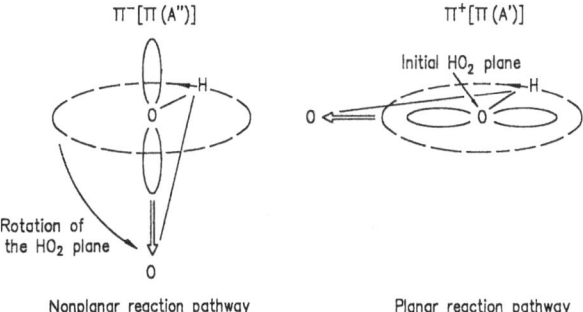

Fig. 7. Vector properties of the $H + O_2 \rightarrow OH\ (N, v, f)$ reaction

well as nonthermal radical concentrations from laser-induced dissociation and isomerization processes can readily be maintained under industrial chemical conditions and used for the production of new, particularly pure products, or products obtained with low energy consumption. Here, the use of the laser offers a number of advantages over conventional light sources: extension of the available wavelength range; the selective excitation of just one kind of molecule, e.g. for isotope separation or ultra-purification because of the narrow spectral bandwidth of the laser light; spatially and temporally controllable, homogeneous excitation of the reaction volume because of the possibilities of pulsed operation and strong collimation; the possibility multiphoton excitation because of the high spectral density. Hg or Xe lamps are mostly used as light sources for industrial photochemistry. Laser photons have to compete both in investment and operational cost as well as in maintenance expenditure, long term power, and lifetime. For economic application of lasers in this area, the effective cost of the photons produced must lie considerably below that for the desired product. One should remember that, generally, the product price is determined only to a small extent by the photochemical step involved in the production and that techniques employing lasers have to compete not only with a conventional photochemical processes, but also with other methods of synthesis. Ideally, the use of lasers should result in several improvements simultaneously, such as cheaper starting materials, fewer or more valuable by-products, and fewer or cheaper process steps. In particular, the production of cheap mass-produced chemicals using lasers is only worthwhile if very high quantum yields (number of product molecules produced per photon generated) can be achieved in radical chain reactions.

Vinyl chloride (VC), the monomer of PVC, is produced industrially mainly by thermal cleavage of HCl from 1,2-dichloroethane (DCE) by a chain reaction. With a world-wide production volume of over 25×10^6 tonnes per annum, VC is one of the leading products of the chemical industry as far as quantity is

Reaction Pathways in the Decomposition of 1,2-dichloroethane

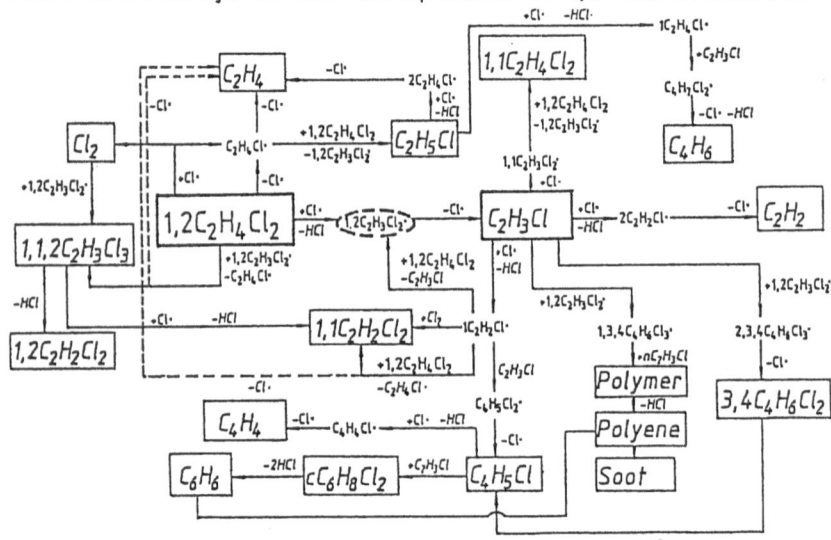

Fig. 8. Reaction paths for the cleavage of hydrogen chloride from dichloroethane, (1, 2-$C_2H_4Cl_2$), in a laser-induced chain reaction

concerned. The advantage of photolytic over thermal initiation of the chain reaction is that the unimolecular process $C_2H_3Cl_2 \rightarrow C_2H_3Cl + Cl$ is rate determining with a low energy barrier. This leads to a low activation energy for the total reaction and hence low reactor temperatures, higher conversions, and fewer by-products [26]. The use of laser radiation to initiate the chain reaction also permits detailed investigations of the reaction kinetics. Radicals may thereby be produced in a wide concentration range and the reactions followed as a function of time. The experimental data can be compared with those obtained from a kinetic model, which for a given temperature and pressure simulates the whole course of reaction (Fig. 8) by using a system of coupled differential equations for the elementary chemical steps. From a comparison of the experimental data with the predictions of the model, missing rate constants of selected elementary steps can be determined [27]. The data so obtained on the pressure and temperature dependence of the laser-induced VC formation can then be used to predict the DCE–VC conversion efficiency for industrially realistic conditions. The calculations show a clear increase in the conversion ratio after laser photocatalysis. This provides a method of using UV-laser radiation to improve a large-scale industrial process. A pilot plant presently under construction is shown schematically in Fig. 9. The thermal cleavage of DCE takes place in a tubular reactor. After heating the DCE, a segment of the reactor can be irradiated with a powerful excimer laser. The laser is tuned to a wavelength at which the absorption in the medium is as small as possible. Due to the highly parallel nature of the laser beam, a very large volume can be irradiated and a steady, small concentration of active chlorine atoms, and hence long

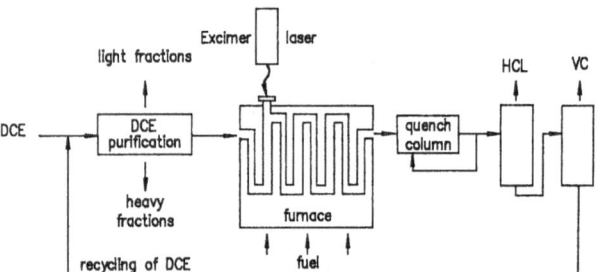

Fig. 9. Pilot plant for the production of vinyl chloride using an excimer laser

chains, are produced. At the same time, irradiation of the wall and hence the initiation of heterogeneous processes is, to a large extent, avoided. With laser radiation, faster conversion at lower temperatures can be achieved compared with the previous process. Thus higher total conversions with simultaneously reduced formation of by-products are possible.

Another way to multiply the effect of laser photons in chemical synthesis is the production of catalysts with laser radiation. By laser pyrolysis of gas mixtures, catalytically active solids of variable composition can be produced (Fig. 10). On complete mixing of the gaseous starting materials and rapid heating in the laser beam, very small solid particles with homogeneous structure and large surface area are obtained, whose composition can be varied over a large range. By optimizing the composition of the laser-synthesized Fe/Si/C catalysts, higher selectivity and preferential formation of the valuable light olefins (C_2–C_4) can be achieved [28]. With the arrangement shown in Fig. 10 fine chrome oxide powders can be produced for the selective catalytic dehydrogenation of saturated hy-

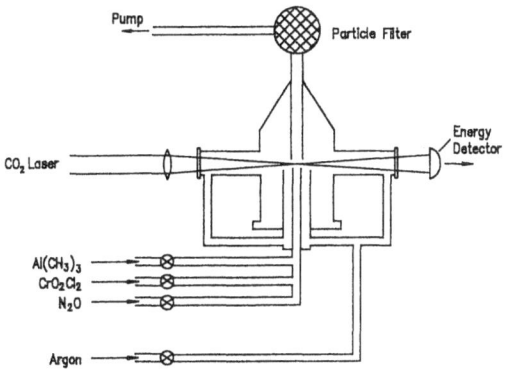

Fig. 10. Synthesis of chromiumoxide catalysts by CO_2 laser pyrolysis of gaseous starting materials. By varying the partial pressures of the component and the laser intensities, catalysts with varying properties can be produced [28, 29]

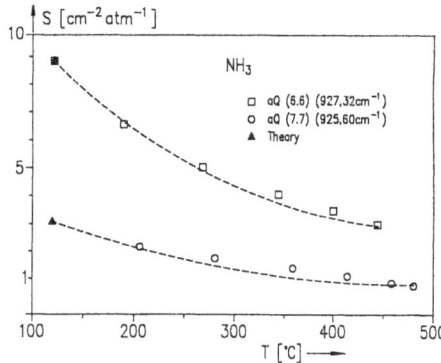

Fig. 11. Temperature dependence of absolute absorption cross sections of NH_3 lines measured with a tunable IR diode laser

drocarbons [29]. Metal carbonyl catalysts can be UV-photoactivated instead of thermally activated. Active centres are formed by cleavage of one or more CO ligands, often resulting in catalytic activity, even at room temperature. Besides polymerization, the isomerization and hydration of alkanes with $Fe(Cl)_5$ catalysts [30, 31] and the silylation of alkanes and aldehyde with $Co_4(CO)_{12}$ catalysts upon irradiation with a XeF laser or a frequency-doubled Nd YAG laser, have been investigated. For the isomerization of 1-pentene, quantum yields of 10^3 and "turn-over" rates (quantum yield per average lifetime of the catalysts) of $4 \times 10^3 \, s^{-1}$ at 50% conversion of 1-pentene can be achieved. Broad-band light sources, such as flash lamps, lead more readily to direct photolysis and to undesired by-products.

3. Laser Spectroscopy of Industrial Chemical Processes

3.1. Application of IR Laser Spectroscopy to In Situ Monitoring of NH_3 in Power Plants

Inhibition and sensitization of chemical processes by nitric oxide are well-known phenomena. Recently, the role of the nitrogen oxides in the formation of acid rain, photochemical smog and the possible depletion of the stratospheric ozone layer has stimulated interest in chemical reactions which can selectively remove nitrogen oxides. An interesting elementary chemical reaction in this respect is the reaction of nitric oxide with the NH_2 radical. First direct studies of the rate and products of this reaction showed a very fast complex addition rearrangement sequence which forms nitrogen molecules and highly vibrational excited water molecules in the single step [32]: $NH_2 + NO \rightarrow N_2 + H_2O^*$. Using time-resolved infrared emission combined with laser photolysis, direct

measurements of the distribution of reaction energy in the water molecule formed in this reaction can be carried out [33]. An enlarged volume for the production of NH_2 radicals by photolysis of NH_3 is created by multi-reflection of an ArF exciplex laser beam combined with a Welsh-mirror light gathering system for the effective collection of infrared fluorescence from excited H_2O molecules. The spectrally resolved infrared emission can be approximately simulated with a vibrational temperature for the H_2O molecules formed of 10^4 K. Significantly lower vibrational temperatures are found in the nitrogen molecules by CARS spectroscopy [33]. The reaction is very selective in channelling the available reaction energy preferably in the stretching vibrations of one reaction product. Therefore, many simultaneous and competing pathways have to be considered in a simulation of the selective reduction of NO by NH_3 in the presence of various amounts of O_2 [34, 35]. Such model calculations can now be applied to realistic situations. Under these conditions one observes a very rapid reduction of NO at the optimum temperature. The model also indicates that rapid mixing of NH_3 with flue gases is essential to obtain optimal reduction results. This can be achieved by using steam or other media for producing a high momentum of the injected material. The main practical problem remaining is the control of the ammonia injection level and location such that breakthrough of NH_3 is avoided under varying combustion conditions. This problem can be attacked by using laser methods for in situ monitoring of NH_3. Figure 11 shows absolute absorption cross sections for NH_3 absorption lines at higher temperatures obtained with a tunable diode laser system. Using $^{13}CO_2$ in an infrared waveguide CO_2-laser, one observes coincidences between NH_3 and $^{13}CO_2$-laser lines in spectral regions where other components of the flue gas (H_2O, $^{12}CO_2$, SO_2, hydrocarbons) do not absorb significantly. The $^{13}CO_2$

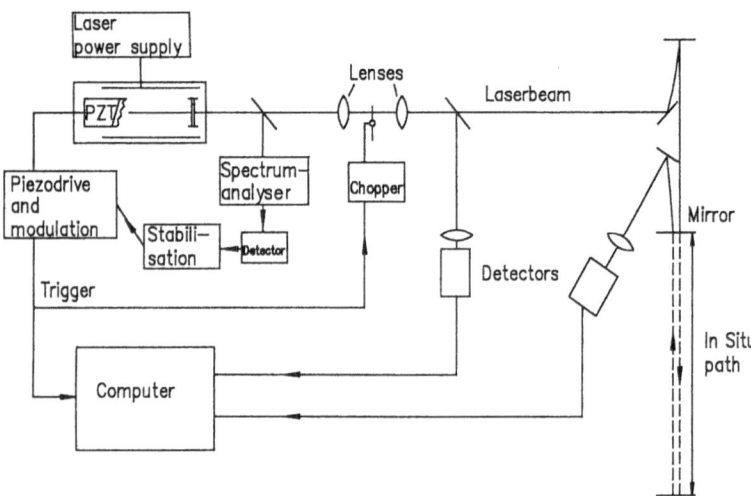

Fig. 12. In situ detection of NH_3 in power
plants using differential absorption

laser lines have a bandwidth of less than $0.01\ \mathrm{cm}^{-1}$ while the distance to the line centre of the NH_3 line can be as little as $0.02\ \mathrm{cm}^{-1}$ where the NH_3 line width is $0.08\ \mathrm{cm}^{-1}$. By choosing a suitable reference line and a multipath absorption arrangement, one can achieve a detection sensitivity of less than 1 vppm for in situ measurements of NH_3 concentrations at higher temperatures by differential absorption [35] with kHz repetition rates (see Fig. 12).

3.2. Non-Intrusive Temperature Measurements in Counter-Flow Diffusion Flames with CARS Spectroscopy

Industrial chemical processes are characterized by a very complex interaction of elementary chemical reactions and various transport processes. Measurements of temperature by thermocouples, and of concentrations of chemical reacting species by probe sampling, can greatly disturb this complicated balance and give results greatly in error. The development of non-intrusive laser methods for monitoring industrial processes opens up new ways to obtain unperturbed information on the temperature and species concentrations which can be compared with detailed mathematical modelling.

Laminar counter-flow diffusion flames constitute an important basis for the simulation of turbulent combustion processes [36]. Having a library of precalculated "flamelets" at different strain rates and fuel/oxidizer mixture compositions, the numerical treatment of turbulent flame structures with a "flamelet model" [37] is reduced to a problem that can be solved within a reasonable amount of time on existing computers. Therefore, there is renewed interest in the study of diffusion flames under a variety of different burning conditions to test and refine the validity of chemistry and transport properties in modelling these flame structures. Among the different types of counter-flow diffusion flames the configuration where the flame is established in the forward stagnation region of a cylindrical porous burner offers some inherent advantages [38]. A schematic drawing of the burner configuration is shown in Fig. 13. A simple nearly 1-dimensional interaction of laminar flow and chemical reaction can be realized in this way. Two beams emitted by a Nd:YAG laser (Quanta Ray DCR1A) at 532 nm were used to pump a broad-band dye laser which delivered about 10 mJ at the Stokes shifted wavelength. The pump beams were achieved by sequentially doubling the fundamental and residual 1.06 µm output from the first frequency doubler. The secondary green beam was used to transversely pump the Stokes dye oscillator. A Brewster plate polarizer ensured an output beam 90% vertically polarized. The dye laser and part of the 532 nm pump beam (30 mJ) were collinearly combined in a USED–CARS phase matching geometry [39]. The CARS signal emerging from the measurement point (actually a cylinder approximately 60 µm in diameter and 1 mm long) was filtered off the residual pump and Stokes beams components and dispersed in a 1.3 m monochromator (McPherson Mod. 209) equipped with a 2400 l/mm holographic grating. The linear dispersion in the plane of the intensified diode array camera (SI, Mod. IRY, 512 pixels) mounted behind a magnifying lens was measured to be $0.14\ \mathrm{cm}^{-1}$ per diode pixel. Throughout the experiments possible non-resonant CARS contributions complicating data evaluation [40] were suppressed by polarization techniques [41]. Averaged spectra (typically 300 laser shots) were stored in a laboratory computer and transferred to a larger computer (IBM 3090–180) for further analysis. Temperatures were deduced from computer-generated least-squares fitting spectral shapes of nitrogen vibrational-rotational Q-branch CARS spectra to measured spec-

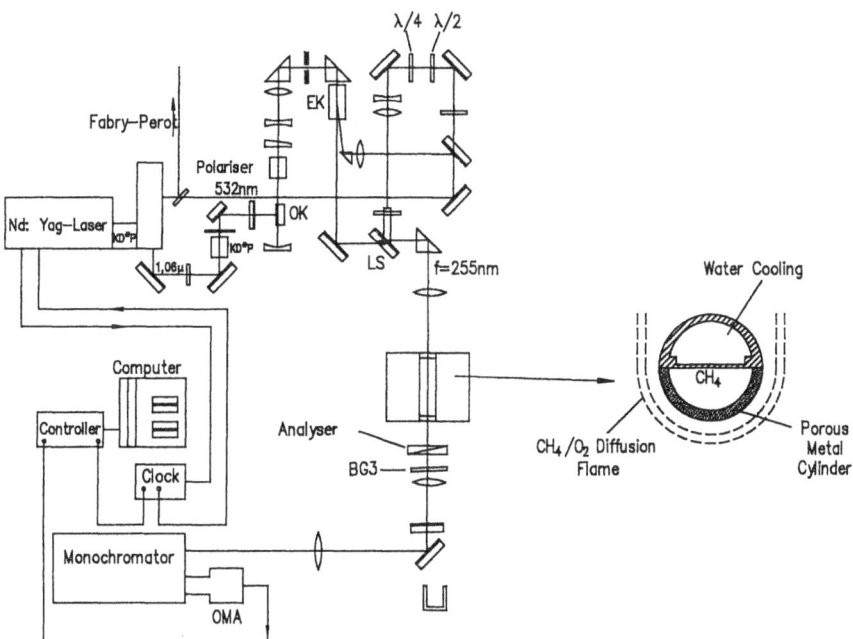

Fig. 13. Experimental
arrangement for measuring
temperature and concentration
profiles in a counter-flow diffusion
flame with broad-band
USED–CARS

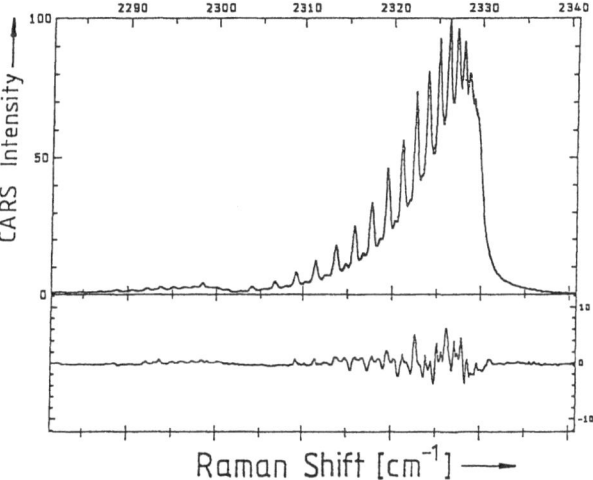

Fig. 14. Nitrogen Q-branch CARS spectrum taken at 2.6 mm
from the burner head in the $a = 200\,\text{s}^{-1}$ counter-flow diffusion
flame. The best fit spectrum, characterized by a temperature of
1870 K, is shown as a dashed line together with the residuals of
both signatures (lower trace in the figure)

tra with temperature as a variable parameter. To get
precise information on temperature from CARS spectra, a simulation program has to take into account
collisional narrowing effects [42]. Another influence
resulted from cross coherence effects induced by using
a multimode laser as a light source in generating CARS
spectra. This phenomenon accounts for partial
coherences in the CARS signals [43, 44]. Cross
coherence effects are mainly induced when correlated
laser fields are used in generating CARS signals leading
to convolution procedures different from the classical
descriptions given by Yuratich [45]. Following a
procedure of Koszykowsky et al. [46], the
temperature-dependent part of the complex G-matrix
was diagonalized. If the off-diagonal elements in the
G-matrix – which describe the relaxation rates between
states of different energy in the colliding molecule and
therefore are responsible for collisional narrowing –
were ignored, the previously used isolated line model
results. In the current version of the fitting program,
routines were included which use the closed form
solution of Greenhalgh and Hall [43]. For the fitting
routine, the procedure commonly used was replaced
by orthogonal "Householder" transformations [47].
To reflect the real measurement situation, the distor-
tion of the spectrum by the detector and spectrograph
instrument functions as well as the pump laser line-
width had to be included in the simulation program by
appropriate convolutions. Best result have been obtained by a priori fitting of these parameters to a CARS
spectrum taken at room temperatures. Figure 14
shows a nitrogen Q-branch CARS spectrum taken in a
counter-flow diffusion flame together with the spec-

trum simulation. In Fig. 15 temperature profiles
measured by CARS in counter-flow diffusion flames
with velocity gradients $a = 2v/r = 250\,\text{s}^{-1}$ and $350\,\text{s}^{-1}$
(v = free stream air velocity, r = radius of the porous
burner) are shown together with the numerical results
[48]. The measured locations of the temperature
maxima agree quite well with those of the calculated
ones. The general shapes and widths of the profiles as
measured with CARS are very similar to those of the
calculated profiles while previous thermocouple measurements [38] show significant discrepancies. The

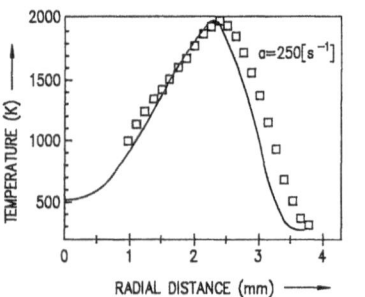 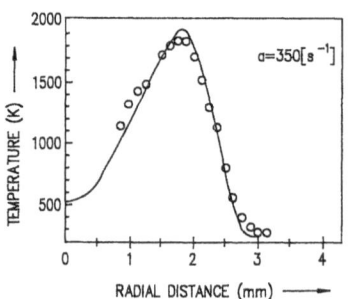

Fig. 15. Temperature profiles calculated (lines) and measured by CARS ($\square$, $\circ$) in CH_4/air counter-flow diffusion flames with different strain rates ($a=2v/r$)

computed peak temperatures in all cases are a little higher than those measured by CARS. This may be partly attributable to the fact that radiative losses are not included in the modelling calculations. Heat flux to the cylinder may be of increasing importance for higher velocity gradients. Similar results are obtained for concentration profiles.

3.3. 2-D Species Concentration Imaging in Turbulent Reactors

The application of planar laser-induced fluorescence (LIF) [49] gives access to multidimensional species

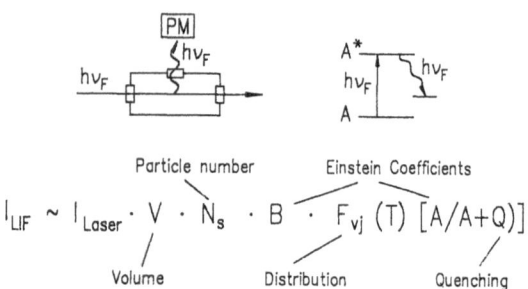

Fig. 16. Principles of laser-induced fluorescence spectroscopy

concentration and temperature and velocity information in reacting flows [49]. As shown in Fig. 16, the intensity of the fluorescence signal can be expressed by the spectroscopic parameters, the ground state population and the electronic quenching rate (Q). Since at atmospheric and higher pressures, fluorescence lifetimes (equivalent to $1/Q$) for many key species in combustion are of the order of nanoseconds or less, an ultra-short-pulse laser coupled with fast detection and data acquisition/processing is required for direct measurements. The hydroxyl radical is an important intermediate in all flames containing hydrogen and oxygen and can serve to indicate the progress of combustion. In laminar atmospheric methane/air flames, for example, concentrations of up to several thousand ppm are generally observed. The laser wavelengths required for excitation are convenient and the fluorescence from the Meinel bands can be used for sensitive detection. Time-resolved measurements were made with a home-built picosecond dye laser (Fig. 17), using an excimer pump laser (308 nm, EMG MSC 103 Lambda Physik). The starting pulse length of 8 ns was shortened first in a quenched transient dye laser (QDTL), which operates by suppressing resonator

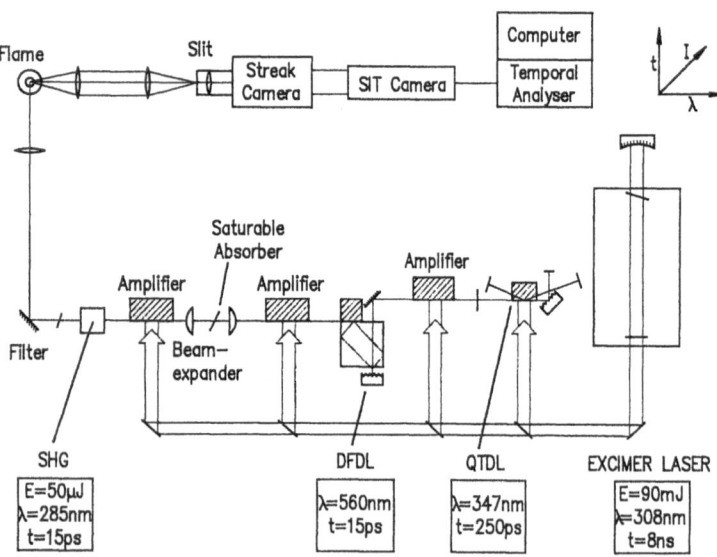

Fig. 17. Experimental set-up for fluorescence lifetime measurements in atmospheric pressure flames. DFDL (distributed feedback dye laser) QTDL (quench transient dye laser) SHG (second harmonic generation)

transients in a double resonator system [50]. After amplification, these pulses were used to pump the oscillator of the second stage, the distributed feedback laser (DFDL) [51]. In the DFDL, feedback is provided by Bragg scattering from spatially periodic perturbations of the optical gain and pulse shortening achieved by spatial separation of the first resonator peak from the following peaks [52]. For OH(A–X) excitation, rhodamine B was used in the DFDL cuvette. The second harmonic of the laser frequency was generated by passing the beam through an 8 mm long crystal of β-BaB$_2$O$_4$. A UV pulse of 15 μJ energy was used with a repetition rate of 1 Hz. The laser beam was directed into an atmospheric pressure, water-cooled (thermostat 50 °C) burner of the flat-flame type. Premixed flows of methane/air were supplied to the burner using Tylan mass flow controllers. Integrated fluorescence from the OH radicals was observed at right angles to the focussed laser beam. Fluorescence light was focussed onto the 100 μm × 15 mm slit of a streak camera (C1587 Hamamatsu, S-20 photocathode), a streak time of 10 ns per 15 mm was set and the data were interpreted by a temporal analyser (Hamamatsu C2200).

For most measurements, the $Q_1 5$ line at 308.52 nm was excited. Measurements of the OH(A–X) lifetime were made at different heights above the burner surface, i.e. from 0.5 to 20 mm, also with a constant height of 20 mm but varying ϕ from 0.77 to 1.43. Within experimental error, no change in lifetime could be discerned in either series of measurements (see Fig. 18). The experimental result, that the collisional quenching rate is constant throughout an atmospheric pressure of flame although for different mixing ratios, allows the determination of absolute number densities in such flames, as shown in Fig. 19. By combining an integrated absorption measurement with spatially resolved LIF [54]. Figure 19b and c shows the OH radical concentration in an unperturbed and perturbed Bunsen burner flame.

Various methods of imaging flame fronts in combustion and flow processes using dopants have been described. Some of these are based on the seeding of particles into the flow to be investigated. For example, TiCl$_4$ added to the combustion chamber reacts with water formed in the flame front to form TiO$_2$ particles, which can serve as centres for Mie scattering and can be used for 2-D visualization. However, such methods involve complex data processing and are not always applicable (e.g. due to particle lag). These limitations can be largely overcome by seeding with fluorescent dopants. Experiments using acetaldehyde as dopant will be described here [56]. Detailed investigations on LIF of acetaldehyde in supersonic jets [57] have provided information on the rovibronic levels of the S_1

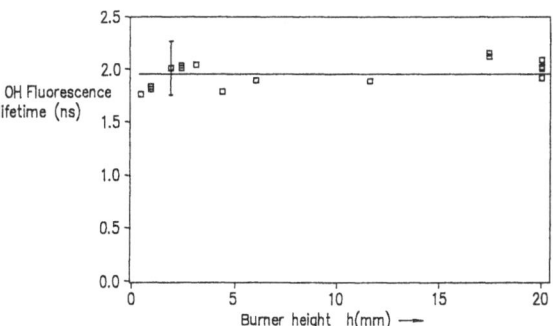

Fig. 18. Lifetime measurements of OH excited at 308.52 nm ($Q_1 5$ line) at various heights in a laminar premixed methane/air flame at atmospheric pressure

state below the photodissociation threshold together with lifetimes for many of these levels, and a more reliable value for the origin of the S_1 state. The conclusion of studies was that the dominant relaxation process from lower vibrational levels of the S_1 is irreversible internal conversion to S_0. Studies of the dependence of acetaldehyde fluorescence on added gas pressures up to 25 bar establish that this molecule is a suitable tracer to show up the flame front in an internal combustion engine using the set-up shown in Fig. 20a. The combustion engine, which was supplied by Daimler-Benz, was operated as a compression expansion machine with a square cross section cylinder to allow easy access for line-of-sight measurements. The XeCl excimer laser light sheet (308 nm) entered the combustion chamber through a quartz window on the cylinder head. The LIF signal from the acetaldehyde in the detection region (35 × 35 mm) was collected by the imaging optics of the image-intensified CCD camera. Stray light was supressed by inserting an interference filter between the imaging optics and the chamber and attaching a dielectric mirror to the top of the piston to reflect the laser beam out of the combustion chamber. Fluorescence images were digitized and processed by a frame grabber (MATROX MUP-AT) and connected to a personal computer (Zenith Z386). Images were then taken at different times after ignition (Fig. 20b), using a detection region of approximately 35 × 35 mm. The bright parts of the image are those where the acetaldehyde/propane/air mixture is still present, in contrast to the dark parts of the image where the fuel has been burnt. The boundary between the burnt and unburnt zones can easily be recognized. Only a small zone with an intermediate concentration of the gas mixture can be seen between the unburnt and burnt regions. The small zone of intermediate fluorescence intensity could not primarily represent a temperature gradient between the flame front and the region of unburnt fuel, since no significant fall in acetaldehyde fluorescence yield with temperature could be observed.

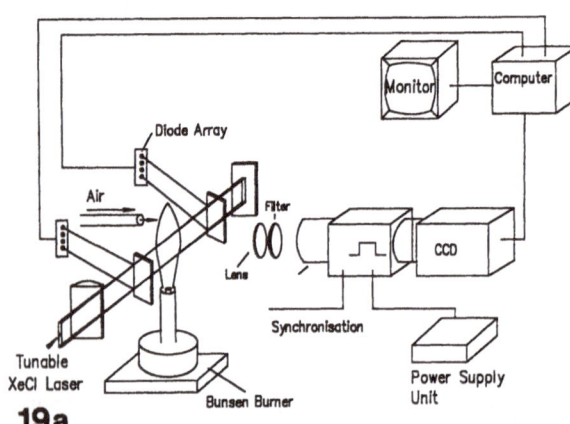

19a

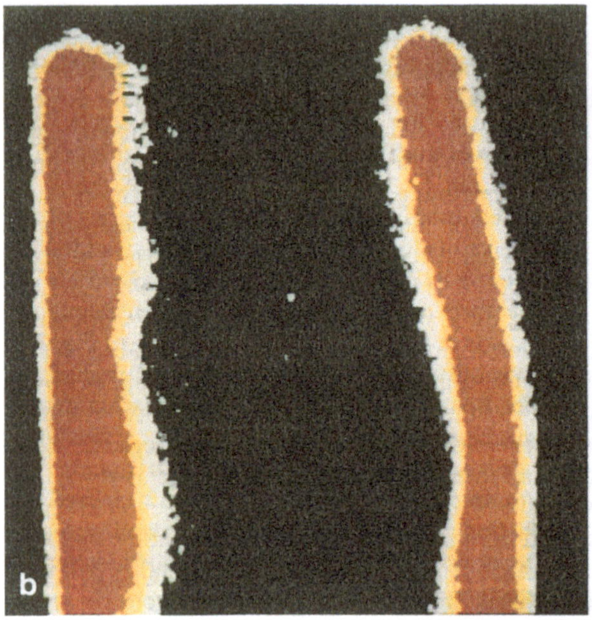

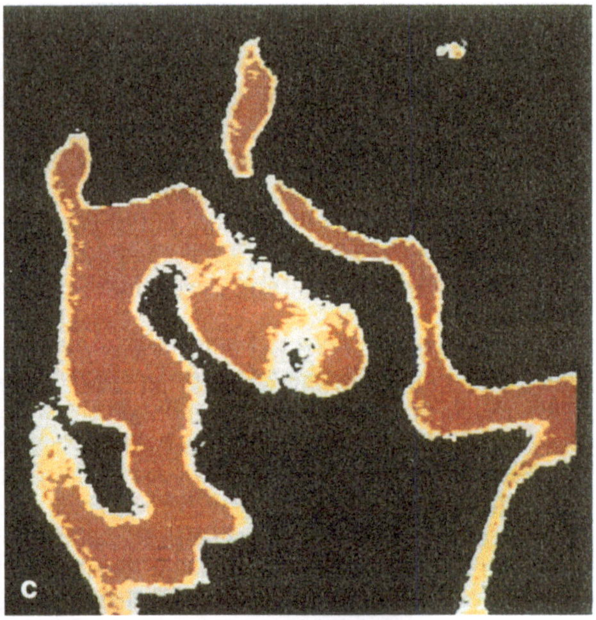

Fig. 19a–c. Determination of 2-*D* instantaneous OH concentrations in flames using tunable excimer lasers. (a) Experimental set-up. (b) OH distribution in an unperturbed Bunsen burner flame. (c) Perturbed flame

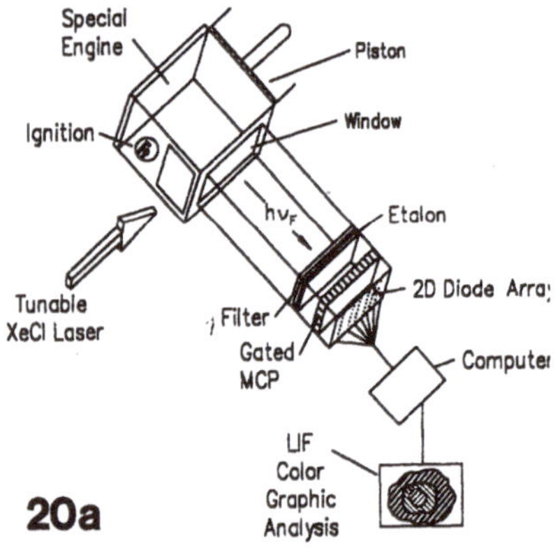

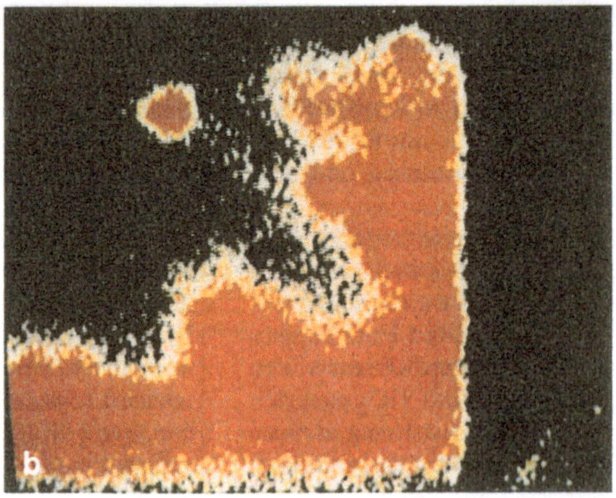

Fig. 20 (a) Experimental set-up for imaging experiments in an internal combustion engine. (b) Imaging of the flame front in an internal combustion engine using acetaldehyde for imaging the flame front in the engine after ignition (burnt parts are the dark regions)

4. Lasers in Genetic Engineering

Many of the spectroscopic techniques described above can also be applied in biochemistry and biology. While biological applications of nonlinear techniques such as CARS or REMPI are still in their infancy, other techniques have contributed valuable information to our knowledge of life processes. For example, practically all information on the primary processes of vision and photosynthesis, i.e. the fast processes where the physical stimulus is transferred into chemical reactions, stems from (sub-)picosecond absorption spectroscopy [58, 59]. Laser-induced fluorescence, due to its high sensitivity, is particularly suited to the study of biological molecules which are available only in small amounts or which are poorly soluble in water, the natural solvent of many biological molecules. A very attractive application of LIF is the study of protein-DNA binding, which is governed by electrostatic interaction of the positively charged amino-acids lysine and arginine with the negatively charged DNA, and by stacking interactions of the aromatic amino-acids tyrosine and tryptophane with the heterocyclic bases of DNA. The aromatic amino-acids reveal intense fluorescence in the ultraviolet. When they bind to DNA, they transfer their fluorescence energy to the latter, which is essentially dark [60]. This fluorescence can therefore be used as a probe of the binding process. LIF experiments have been used to elucidate the binding kinetics of peptides such as lysine-tyrosine-lysine to DNA. These peptides first bind unspecifically via electrostatic bonds and in a second step specifically (i.e. recognizing topological features of the DNA) via stacking [61]. Such a stacking interaction also seems to be responsible for the stabilization of a simple bacterial virus, the phage Pf1, and for an unusually large base-to-base distance in a rare structural form of DNA (inverted DNA) [62].

LIF experiments have also been used to study the electrostatic part of protein-DNA interaction. While lysine forms pure electrostatic bonds [63], LIF has revealed that arginine interacts via a mixture of electrostatic and non-electrostatic bonds [64]. This difference may explain why, in the life of a biological cell, arginine-rich proteins are occasionally replaced by lysine-rich proteins.

Studies of the wavelength dependence of laser interactions with the surface of human red blood cells have shown that UV photons interact much more effectively with biological cells than visible or IR light [65]. This finding has led to the extension of laser microbeam techniques [66] to the ultraviolet.

A UV laser microbeam can be realized by an excimer-laser-pumped dye laser, the pulses of which are coupled into a microscope and focussed to the diffraction limit. At the focal point, energy densities of 10^{10}–10^{12} W/cm^2 can be obtained. These energies are sufficient to manipulate biological objects with spatial accuracy of the wavelength of the light used, and even below that, if nonlinear effects can be exploited. Spatial accuracy is not only obtained in the object plane but also along the optical axis. Because energy densities sufficient for manipulation of biological material are obtained only a few hundred nanometres above and below the focal point, one can manipulate subcellular

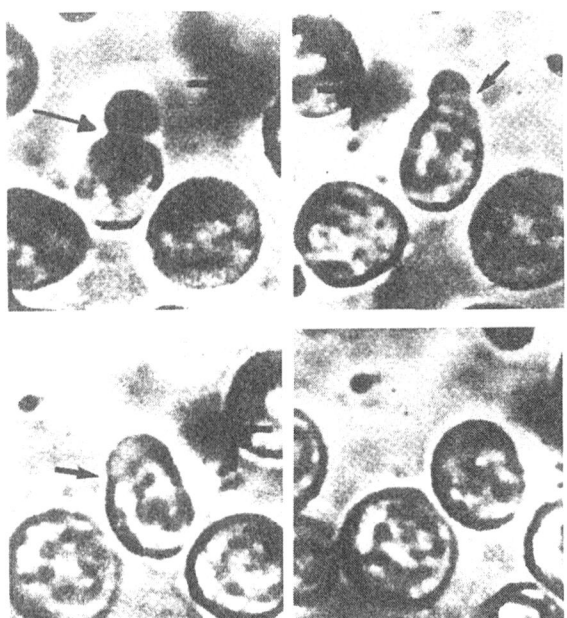

Fig. 21. Fusion of a mouse *B*-lymphocyte with a mouse myeloma cell

structures in a cell without opening it. This UV laser microbeam can thus be used for the treatment of biological cells and subcellular structures. For example, individual cells can be fused under total microscopic control [67, 68], genetic material can be introduced into plant cells [69] and chromosomes can be dissected into small portions [70].

Figure 21 gives a series of photographs of a cell fusion process, taken from a microscope at a magnification of 1000. The small cell is a *B*-lymphocyte, a cell from the immune system of a mouse, capable of producing antibodies. In vitro *B* lymphocytes die after 3–4 generations, i.e. they cannot be held in culture. The large cell is a long-lived myeloma (blood cancer) cell. Both cells are induced to fuse by a few UV pulses focussed on the contact area. The major advantages of this laser-induced cell fusion, as compared with conventional cell fusion techniques, are the total microscopic control and the fact, that the process occurs under physiological conditions. Therefore, even fragile cells have a chance of surviving the fusion process. By combining laser-induced fusion with coupling of the two fusion partners via an antigen-avidin-biotin bridge [72], one can preselect those *B*-lymphocytes which have the potential to produce antibodies against a given antigen. Only those *B*-lymphocytes are fused with myeloma cells to give hydridoma, the long-lived fusion products with the potential to produce antibodies in vitro. Based on this technique, a method is being developed which avoids the lengthy search for the hybridoma specific for the "wanted" antibody. If

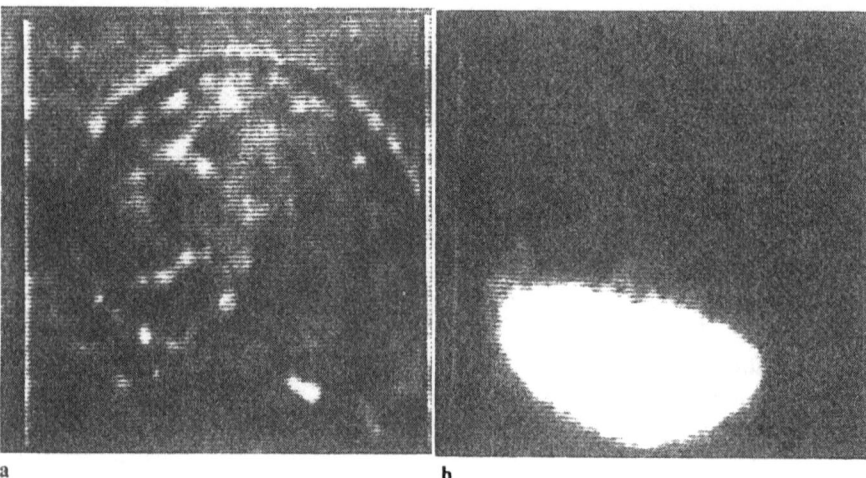

a   b

Fig. 22a, b. Injection of
fluorescently labelled DNA into a
rape cell using an UV laser
microbeam

some remaining cell-biological problems can be solved, this may become the basis for a fast, specific production of monoclonal antibodies for cancer therapy. Such a fast method is required since cancer cells change their susceptibility to a special antibody within weeks, while present methods for the production of antibodies take months to years.

Figure 22a shows a plant cell in which foreign DNA has been injected by use of the UV laser microbeam. A self healing hole of 0.5 to 1 µm is punched into the cell wall of the plant cell and fluorescently labelled DNA is forced into the cell by slight osmotic pressure differences. The uptake of the DNA is imaged by the fluorescence in a part of the cell in Fig. 22b. This direct transfer of DNA into plant cells is applicable for all cell types, and thus particularly crop plants can be genetically modified, in which other DNA transfer methods are only occasionally successful.

The possibility to work in the depth of a cell without opening it allows the study of the function of subcellular particles. For example, the mitotic spindle, the subcellular system which is responsible for accurate separation of chromosomes during cell division, can be inactivated by microbeam irradiation [66]. Since erroneous chromosome partition may be responsible for diseases such as mongolism (where chromosome 21 occurs in three instead of two copies), the laser microbeam may become an attractive tool to study the basic mechanism of these diseases. Furthermore, this technique is suitable for studying the mechanism of intracellular transport processes. For example, the cytoplasmic streaming is stopped by focussing a UV laser microbeam into the interior of a pollen tube and calcium is released from depots into the cytoplasm [71]. Studies are currently in progress to reveal the role of this calcium release in intracellular transport processes.

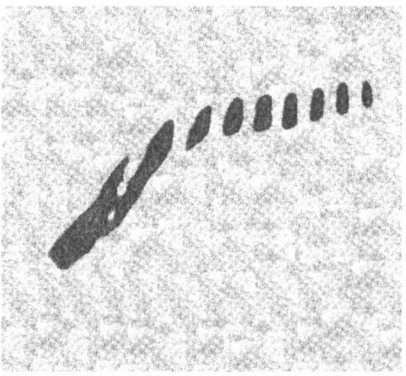

Fig. 23. Human chromosome dissected into slices of 300 nm using the diffraction rings of a laser microbeam

Figure 23 shows a human chromosome which is dissected into equal slices. For this experiment, diffraction, which is usually considered as an unwanted side effect of focussing to the theoretical limit, has been used to microdissect the chromosome. Microdissection is a basic technique used to study the molecular basis of disease. Many diseases such as muscular dystrophy, Alzheimer, cystic fibrosis, leukaemia and solid tumours are correlated with a defect in a specific region of a chromosome. In the past, the study of the molecular basis of such a disease took of the order of ten years, as in the case of the recently solved muscular dystrophy [73]. In those studies the portion of the X chromosome, which is defective in the disease, was isolated by sophisticated cell-biological methods. Laser microdissection promises to speed up that process by at least one order of magnitude, as is indicated in studies on the molecular basis of cystic fibrosis which are presently under way. With chromosomal material from the fruit fly drosophila, it has already been shown that the molecular-biological techniques required for molecular analysis of disease can be applied to chromosome slices after laser microdissection [74].

Acknowledgements. Support of this work by the BMFT (TECFLAM, LABIO), DFG (SFB 123), Stiftung Volkswagenwerk, Fonds der Chemischen Industrie and the Commission of the European Communities is gratefully acknowledged. I also thank Dr. K. O. Greulich for compiling the section on biological applications of lasers.

References

1. F.P. Schäfer, W. Schmidt, J. Volze: Appl. Phys. Lett. **9**, 306 (1966)
   P.P. Sorokin, J.R. Lankard: IBM J. Res. Dev. **10**, 162 (1966)
2. F.P. Schäfer: Appl. Phys. B **46**, 197–206 (1988)
3. S. Arrhenius: Z. Physik. Chem. **4**, 226 (1889)
4. F. London: Z. Elektrochem. Angew. Phys. Chem. **35**, 592 (1929)
   P. Siegbahn, B. Liu: J. Chem. Phys. **68**, 2455 (1978); **80**, 581 (1984)
5. J. Arnold, D. Chandler, Th. Dreier: J. Chem. Phys. (to be published)
6. W. Meier, G. Ahlers, H. Zacharias: J. Chem. Phys. **85**, 2599 (1986)
7. Th. Dreier, J. Wolfrum: Int. J. Chem. Kinet. **18**, 919 (1986)
8. J. Wolfrum: Disc. Faraday Soc. **84** (1987)
9. U. Wellhausen, J. Wolfrum: Ber. Bunsenges. Phys. Chem. **89**, 314 (1985)
10. M. Kneba, J. Wolfrum: J. Phys. Chem. **83**, 69 (1979)
    D.K. Bondi, J.N.L. Connor, J. Manz, J. Römelt: Mol. Phys. **50**, 467 (1983)
11. K. Kleinermanns, J. Wolfrum: Angew. Chem. Int. Ed. Egl. **26**, 38 (1987)
12. D. Arnoldi, K. Kaufmann, J. Wolfrum: Phys. Rev. Lett. **34**, 1597 (1975)
13. J. Warnatz: Ber. Bunsenges. Phys. Chem. **87**, 1008 (1983)
    J. Wolfrum: 20th Symp. (Int.) on Combustion, The Combustion Institute (1984) p. 559
14. K. Kleinermanns, J. Wolfrum: J. Chem. Phys. **80**, 1446 (1984)
15. K. Kleinermanns, R. Schinke: J. Chem. Phys. **80**, 1440 (1984)
16. K. Kleinermanns, E. Linnebach: Appl. Phys. B **36**, 203 (1985)
17. I. Schechter, R.B. Bernstein, R.D. Levine: J. Phys. Chem. **91**, 5466 (1987)
18. J.A. Miller: J. Chem. Phys. **74**, 5120 (1981)
19. C.F. Melius, R.J. Blint: Chem. Phys. Lett. **64**, 183 (1979)
20. C. Cobos, H. Hippler, J. Troe: J. Phys. Chem. **89**, 342 (1985)
    J. Troe: J. Phys. Chem. **90**, 3485 (1986)
    J. Troe: Combust. Flame (1988) (in press)
21. T.H. Dunning, Jr., S.P. Walch, M.M. Goodgame: J. Chem. Phys. **74**, 3482 (1981)
22. G.J. Vazquez, S.G. Peyerimhoff, R.J. Buenker: Chem. Phys. **99**, 239 (1985)
23. G.L. Schott: Combust. Flame **21**, 357 (1973)
24. Th. Just, P. Frank: Ber. Bunsenges. Phys. Chem. **89**, 181 (1985)
25. A.N. Pirraglia, J.V. Michael, J.W. Sutherland, R.B. Klemm: J. Phys. Chem. (1988) (in press)
26. P.N. Clough, M. Kneba, M. Schneider, J. Wolfrum: Europäische Patentanmeldung Nr. 801056.490
27. M. Schneider, J. Wolfrum: Ber. Bunsenges. Phys. Chem. **90**, 1058 (1986)
28. A. Gupta, J.T. Yardley: Proc. SPIE Int. Soc. Opt. Eng. **458**, 131 (1984)
29. J. Kern, H. Schwahn, B. Schramm: Mat. Chem. Phys. (1988)
30. R.L. Whetten, K.J. Fu, E.R. Grant: J. Am. Chem. Soc. **104**, 4270 (1982)
31. J.C. Mitchener, M.S. Wrighton: J. Am. Chem. Soc. **103**, 975 (1981)
32. M. Gehring, K. Hoyermann, H. Schacke, J. Wolfrum: 14th Symp. (Int.) on Combustion, The Combustion Institute (1973) p. 99
33. Th. Dreier, J. Wolfrum: 20th Symp. (Int.) on Combustion, The Combustion Institute (1984) p. 695
34. R.K. Lyon: U. S. Patent Nr. 39.00.55.4
35. H. Hemberger, H. Neckel, J. Wolfrum: Third TECFLAM Seminar. Stuttgart (1987) p. 47
36. G. Damköhler: Z. Elektrochem. **46**, 601 (1940)
37. K.N.C. Bray, P.A. Libby: Phys. Fluids **19**, 1687 (1976)
38. H. Tsuji: Progr. Energy Combust. Sci. **8**, 93 (1982)
39. A.C. Eckbreth, G.M. Pobbs, J.H. Stufflebeam, P.A. Tellex: Appl. Opt. **23**, 1328 (1984)
40. R.J. Hall, L.R. Boedeker: Appl. Opt. **23**, 1340 (1984)
41. L.A. Rahn, L.J. Zych, F.L. Mattern: Opt. Commun. **30**, 249 (1979)
42. J.R. Hall, J.F. Verdieck, A.C. Eckbreth: Opt. Commun. **35**, 69 (1980)
43. D.A. Greenhalgh, R.J. Hall: Opt. Commun. **57**, 125 (1986)
44. R.E. Teets: Opt. Lett. **19**, 226 (1984)
45. M.A. Yuratich: Mol. Phys. **38**, 625 (1979)
46. M.H. Koszykowsky, R.L. Farrow, R.E. Palmer: Opt. Lett. **10**, 478 (1985)
47. A. Kim: J. Chem. Education **47**, 120 (1970)
48. G. Dixon-Lewis, T. David, P.H. Gaskell, S. Fukutani, H. Jinno, J.A. Miller, R.J. Kee, M.D. Smooke, N. Peters, E. Effelsberg, J. Warnatz, F. Behrendt: 20th Symp. (Int.) on Combustion, The Combustion Institute (1984) p. 1893
    Th. Dreier, B. Lange, J. Wolfrum, M. Zahn, F. Behrendt, J. Warnatz: Ber. Bunsenges. Phys. Chem. **90**, 1010 (1986)
    E. Dießel, Th. Dreier, B. Lange, J. Wolfrum, F. Behrendt, J. Warnatz: 22nd Symp. (Int.) on Combustion, The Combustion Institute (1988)
49. R.K. Hanson: 21st Symp. (Int.) on Combustion, The Combustion Institute (in press)
50. Z. Bor, B. Racz: Appl. Opt. **24**, 1910 (1985)
    F.P. Schäfer: Laser Optoelektron. **2**, 95 (1984)
51. Z. Bor: IEEE J. QE-16, 517 (1980)
    A. Müller, Z. Bor: Laser Optoelektron. **3**, 187 (1984)
52. S. Szatmari, Z. Bor: Appl. Phys. B **34**, 29 (1984)
53. R. Schwarzwald, P. Monkhouse, J. Wolfrum: Chem. Phys. Lett. **142**, 15 (1987)
54. H.M. Hertz, M. Alden: Appl. Phys. B **42**, 97 (1987)
55. L.D. Chen, W.M. Roquemore: Combust. Flame **66**, 81 (1986)
56. H. Becker, K. Kleinermanns, P. Monkhouse, R. Suntz, J. Wolfrum, J. Koehler, G. Ziegler: 22nd Symp. (Int.) on Combustion, The Combustion Institute (1988)
57. M. Noble, E.K.C. Lee: J. Chem. Phys. **80**, 134 (1984)
58. H.J. Polland, M.A. Franz, W. Zinth, W. Kaiser, E. Kölling, D. Oesterhelt: Biophys. J. **49**, 657 (1986)
59. J.E. Rudzki, K.S. Peters: Biochemistry **23**, 3843 (1984)
60. L. Stryer: Ann. Rev. Bioch. **47**, 819 (1978)
61. C. Helene, J.C. Maurizot: Crit. Rev. Bioch. **10**, 213–258 (1981)
62. K.O. Greulich, R.W. Wijnaendts van Resandt: Bioch. Bioph. Acta **782**, 446 (1984)
63. M.T. Record, C.F. Anderson, T.H. Lohmann: Q. Rev. Bioph. **11**, 103 (1978)

64. J. Ausio, K.O. Greulich, E. Haas, E. Wachtel: Biopolymers **23**, 2559 (1984)

65. K.O. Greulich, S. Monajembashi, T. Cremer, C. Cremer, P. Butz, J. Wolfrum: Proceedings of the CLEO, San Francisco (1986) p. 72

66. M.W. Berns, J. Aist, J. Edwards, K. Strahs, J. Girton, P. Mc Neill, J.B. Rattner, M. Kitzes, M. Hammer-Wilson, L.H. Liaw, A. Siemens, M. Koonce, S. Peterson, S. Brenner, J. Burt, R. Walter, P.J. Bryant, D.A. Van Dyk, J. Coulombe, T. Cahill, G.S. Berns: Science **213**, 505 (1981)

67. R. Wiegand, K. Zimmermann, S. Monajembashi, H. Schäfer, G.M. Hänsch, K.O. Greulich, J. Wolfrum: Immunobiology **173**, 320 (1986)

68. R. Wiegand, G. Weber, K. Zimmermann, S. Monajembashi, J. Wolfrum, K.O. Greulich: J. Cell Sci. **88**, 145 (1987)

69. G. Weber, S. Monajembashi, K.O. Greulich, J. Wolfrum: Naturwissenschaften **75**, 35 (1988)

70. S. Monajembashi, C. Cremer, T. Cremer, J. Wolfrum, K.O. Greulich: Exp. Cell. Res. **167**, 262 (1986)

71. H.D. Reiss, K.O. Greulich, J. Wolfrum, K. Zimmermann: Eur. J. Cell. Biol. (1988) (in press)

72. M.M.S. Lo, T.Y. Tsong, M.K. Konrad, S.M. Strittmater, L.D. Hesler, S.H. Snyder: Nature **310**, 792 (1984)

73. E.P. Hoffmann, R.H. Brown, Jr., L.M. Kunkel: Cell. **57**, 919 (1987)

74. N. Ponelies, E.K.F. Bautz, S. Monajembashi, J. Wolfrum, K.O. Greulich: Chromosoma (to be published) (1988)

Appl. Phys. B 46, 237–251 (1988)

Laser-Induced Chemistry –
Basic Nonlinear Processes and Applications

V. S. Letokhov

Institute of Spectroscopy, USSR Academy of Sciences, USSR,
SU-142092 Moscow Region, Troitzk, USSR

Received 29 February 1988/Accepted 7 March 1988

Abstract. Many methods and achievements in chemistry are based on using the interaction of light with atoms and molecules. It is sufficient to mention photochemistry, flash-photolysis, spectrochemistry and others. The advent of laser amplified the connection between chemistry and light. Today laser light has become a very versatile and effective tool, first, to study the dynamics of chemical reactions, secondly, to stimulate chemical reactions and finally, to analyze substance. The unique properties of laser light (high power, monochromaticity, short duration, directivity and temporal coherence) provide quite new instrumental possibilities in all these problems.

PACS: 32, 42, 82.50

General Principles

1.1. Stimulated Transitions

The laser, however, is not only important for chemistry as an instrument but also gives rise to "laser ideology" in action of light on matter. Indeed, the laser employs stimulated quantum transition from upper quantum levels to the lower ones followed by stimulated emission and *extraction* of coherent light energy from the excited matter (Fig. 1a). On the other hand, the intense laser light being in resonance with quantum transitions of the substance excites the high-lying levels due to successive stimulated upward transitions, i.e. *deposits* the coherent light energy in the substance (Fig. 1b). Stimulated resonant transitions in matter under the action of laser light have very important properties for applications in chemistry. First, the rate of excitation due to stimulated transitions $m \leftarrow n$ is proportional to the light intensity I

$$W_{exc}^{nm} = \sigma_{nm} I(\omega_{mn}),\tag{1}$$

where σ_{nm}, ω_{nm} are the cross-section and frequency of the resonant transition between the quantum levels "m" and "n". This allows a *very high rate* of energy deposition into an atom or a molecule (1–10 eV during 10^{-8}–10^{-12} s, i.e. up to 10^{13} eV/s) that exceeds greatly

the relaxation rate W_{relax}. On condition that $W_{exc} \gg W_{relax}$ it is possible to excite considerably the internal degrees of freedom (vibrational and/or electronic) of atom or molecule without any heating. Secondly, the *resonant* character of stimulated transitions enables exciting atoms or molecules of a definite sort in mixtures, that is, providing intermolecular excitation selectivity. Thus, laser resonant excitation prepares a substance in a highly-nonequilibrium state with relation to both different degrees of freedom and particles of different sorts in a mixture.

1.2. Types of Photoexcitation

The methods of laser-induced chemistry can be classified according to the type of photoexcitation. All methods of single-photon linear photochemistry are based on the excitation of the electronic state of the atom or molecules (single-photon electronic photochemistry) or the vibrational state of the molecule (single-photon vibrational photochemistry) as one photon is absorbed (Fig. 2a and b). The excitation of atomic and molecular states by visible UV irradiation of ordinary sources is well known in photochemistry [1, 2]. Laser radiation here turns out to be very useful for two reasons. First, the high spectral brightness of

STIMULATED QUANTUM TRANSITIONS

in LASERS   in LASER CHEMISTRY

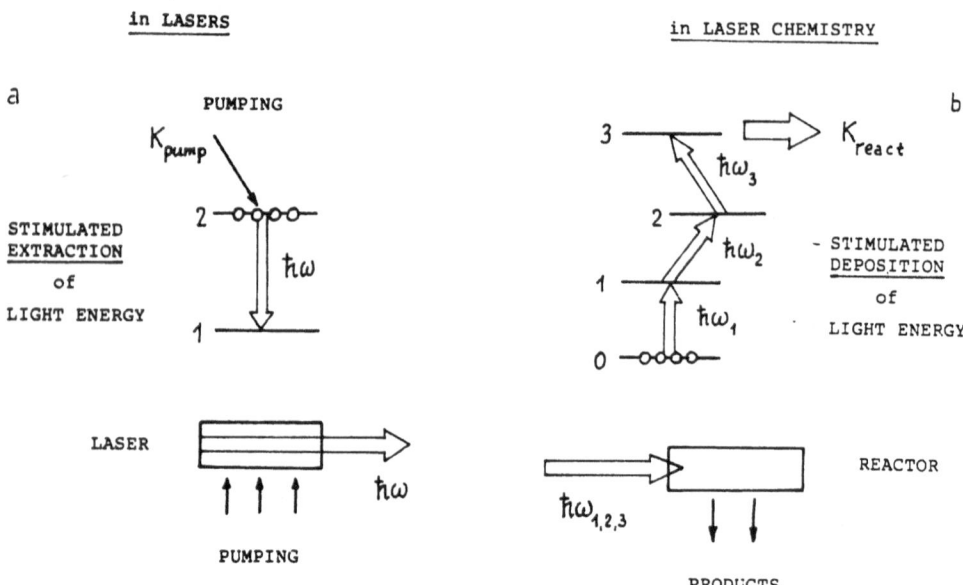

Fig. 1a, b. Laser utilizes the stimulated quantum transitions from an excited level to a lower level for the extraction of stored energy as the coherent light beam (a); the stimulated quantum transitions from low-lying levels to upper levels are utilized for deposition of laser light energy in atoms and molecules (b)

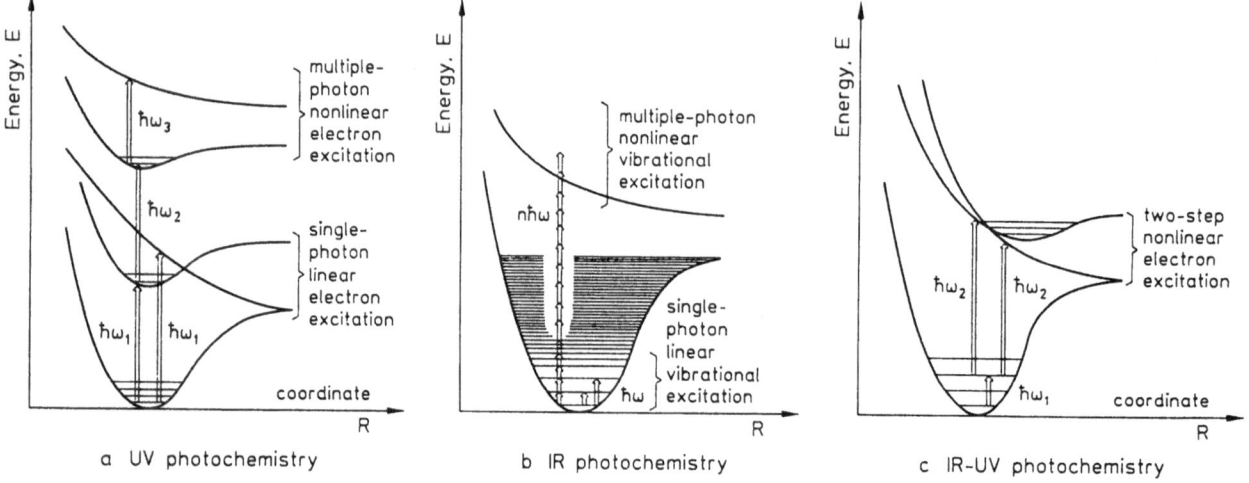

a UV photochemistry   b IR photochemistry   c IR-UV photochemistry

Fig. 2a–c. Various schemes of single-photon (linear) and multiple-photon (nonlinear) excitation of vibrational or electronic levels of a molecule by laser radiation which forms the basis for UV (a), IR (b), and IR-UV (c) photochemistry

laser light makes it possible to excite any discrete quantum state of an atom or molecule without accidental coincidences between the wavelengths of the intense lines of spontaneour radiation of ordinary sources and the absorption lines. Second, the high spectral brightness of IR lasers permits the excitation of molecular vibrations.

The high intensity of laser radiation has allowed new methods of multiphoton (MP), nonlinear photochemistry to be developed, which have no analogy in conventional photochemistry. These methods are

based on absorption of a number of laser photons by one particle (an atom or a molecule). But first, techniques of multistep excitation [3,4] of atomic and molecular electronic states had to be developed (Fig. 2a and c). These methods were based on the ability of laser light to transfer a considerable fraction of atoms and molecules from the ground state to the excited state when the absorption of photons by excited particles becomes probable. The development of high-intensity IR lasers made it possible to realize [5,6] collisionless multiple-photon excitation of high-lying vibrational levels of polyatomic molecules by intense IR radiation (10^6–10^8 W/cm^2) as shown in Fig. 2b. Methods of multiphoton vibrational photochemistry developed rapidly after the discovery [7] of isotope-selective photodissociation of molecules in intense IR fields, and they are now used for isotope separation.

Thus, with modern lasers it is possible to excite the electronic and vibrational levels of molecules. These schemes can be subdivided into electronic (UV) photochemistry with one-photon and multiphoton excitation (Fig. 2a), vibrational (IR) photochemistry with one-photon and multiple-photon excitation (Fig. 2b), and combined vibrational-electronic (IR-UV) photochemistry with multistep excitation (Fig. 2c). In the first two cases the laser radiation interacts either with the electronic or with the vibrational motion of the molecule. Such a subdivision is not strictly due to the inevitable coupling of electronic and vibrational motions. In the final case (Fig. 2c), laser IR and UV radiation excites both the vibrational and the electronic motions of molecules. All these methods of nonlinear laser photochemsitry are reviewed in monography [8].

1.3. Selectivity of Photoexcitation

In laser chemistry an important feature of each different approach is the "selectivity" it offers, which, of course, determines the kind of work it is suitable for. As an example, with which we can discuss different kinds of selectivity, let us consider IR MP laser chemistry.

Resonant MP excitation of molecular vibrations in an intense IR field forms the basis for several essentially different approaches to laser chemistry. They can be classified according to the relations between the various relaxation times for an excited vibrational level interacting with the IR field:

$$\tau_{stoch} \ll \tau_{transf} \ll \tau_{relax}. \qquad (2)$$

Here τ_{stoch} is the time for intramolecular vibrational redistribution (IVR) of absorbed energy between different vibrational modes of molecules being excited; τ_{transf} is the time for intermolecular transfer of vibrational energy between molecules of different kinds in a mixture (for example, molecules of different isotopic composition), and τ_{relax} is the relaxation time for molecular vibrational energy to be transferred to translational degree of freedom – which is the time for complete thermal equilibrium to be reached in the molecular mixture.

The rate of vibrational excitation of a molecule by multiphoton absorption W_{exc}, according to (1), depends on radiation intensity and vibrational transition cross sections. From a consideration of inequality (2) we can distinguish four different conditions for the relaxations between W_{exc} and the relaxation rates of vibrational energy in different circumstances. Accordingly, there are four different approaches to IR MP laser photochemistry depending on how far from equilibrium the vibrational excitation in the molecule and molecular mixture lies (Fig. 3): 1) mode-selective photochemistry; 2) intermolecular selective photochemistry; 3) nonequilibrium chemistry; 4) thermal chemistry. All above-mentioned methods of laser chemistry can be applied in industrial chemistry, though the degree of development of various applications is quite different. IR MP and IR-UV MP

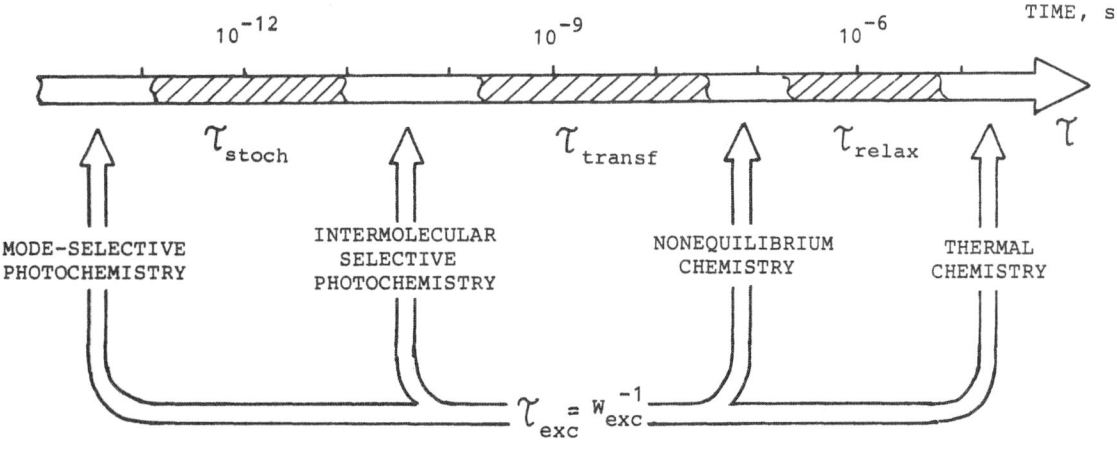

Fig. 3. Various rates of laser photoexcitation or deposition of light energy to the molecule can provide various types of laser-induced chemistry: from future mode-selective photochemistry for very fast subpicosecond rate of excitation to ordinary thermal chemistry for low-rate millisecond excitation. Real values of relaxation times depend on the density of the irradiated substance

selective photochemistry for *isotope separation* is the most advanced application. The selective generation of specific radicals in a molecular mixture by UV one-photon or IR MP photolysis has applications for laser radical *chemical synthesis*. Radicals generated by IR photolysis are very useful for chemical-induced processes at a surface, in particular, for *surface etching* of semiconductor materials. Finally, the laser excitation of high-lying electronic states of organic molecules can be useful for *bio-organic* chemical synthesis. All of these listed trends of laser photochemistry having great potential for industrial applications are reviewed in lectures at this Symposium.

2. Multiphoton Vibrational Photochemistry

The intense IR radiation of a pulsed laser at a frequency tuned to the vibrational absorption band of any polyatomic molecule induces very fast vibrational excitation of that molecule [9]. In other words, under a resonant IR field the vibrational degrees of freedom are subjected to strong heating which depends on the number of IR photons absorbed by the molecule. Under the action of an IR pulse with its energy fluence varying from 1 to $10 \, J/cm^2$ it is relatively easy to deposit energy of 1–10 eV into a polyatomic molecule via the absorption of, say, 10–100 photons from a CO_2 laser operating in the region of $10 \, \mu m$. For such a strong vibrational excitation the rate of photoexcitation W_{exc} and the laser pulse duration τ_p must satisfy the following requirements:

$$W_{exc} = \bar{\sigma}_{abs} I \gtrsim \tau_p^{-1} \gtrsim W_{relax}, \qquad (3)$$

where $\bar{\sigma}_{abs}$ is the average cross-section of successive vibrational upward transitions of the molecule, W_{relax} is the relaxation rate of vibrational states. This condition of efficient excitation is valid when the laser pulse energy fluence $\Phi = I\tau_p \gtrsim \sigma_{abs}^{-1}$ (photo/cm^2). For effective absorption of n photons by the single molecule the energy fluence Φ of laser pulse evidently should satisfy the more stringent condition: $\Phi \gtrsim n/\sigma_{abs}$ (photons/cm^2). The IR MP absorption cross sections for different polyatomic molecules lie in the range $\sigma_{abs} = 10^{-18} - 10^{-20} \, cm^2$.

The qualitative model of the MP vibrational excitation and dissociation of an isolated molecule under the action of an intense IR radiation pulse was developed around a decade ago [10–12]. According to the model, the molecule in the course of its vibrational excitation passes through the following three qualitatively different vibrational energy regions: 1) the region of low-lying discrete vibrational-rotational levels wherein a single transition is possible from a given state, 2) the *vibrational quasicontinuum* (QC) region wherein many close vibrational levels interact and

many transitions are possible from a given state, and 3) the real continuum region above the dissociation limit wherein the unimolecular decay of the vibrationally overexcited isolated molecule becomes possible. At the same time, a deeper insight was gained in the past decade into the MP excitation process, and its more detailed model, illustrated in Fig. 4, was formed as a result of the development of the simple model described above. The most essential addition here is the understanding of how the IR absorption spectrum of the vibrational mode in resonance with the exciting IR radiation evolves with the increasing vibrational energy of the molecule as a whole.

The molecule absorbs the first few IR photons as usual on the quantum transition between the lower discrete vibrational-rotational levels. The intramode anharmonicity of the resonant vibrational mode in this region gives rise to a red shift of the IR absorption band centre. The interaction of the exciting IR field with molecular vibrations has a resonant or mode-selective, but not necessarily coherent, character. The interaction of the modes themselves for the present does not cause their intermixing, unless there is a casual coincidence of the frequencies of several modes, which gives rise to the so-called *intermode resonances*. Such intermixing becomes quite inevitable when the vibrational energy of the molecule becomes high enough. Of essence here is the fact that a polyatomic molecule has many vibrational degrees of freedom. Even a weak intermode anharmonicity is therefore sufficient for many intermode resonances, otherwise known as the Fermi resonances, to occur. The overlapping of the intermode resonances provides for the stochastization of the molecular vibrational energy [13]. As the Fermi resonances are approximate, i.e., take place at frequency detunings of tens of cm^{-1}, to make them overlap at a given intermode anharmonicity, it is necessary that their density should be sufficiently high, and this is possible only at a certain vibrational energy store. For this reason, the *stochastization* of vibrational energy starts from some threshold energy, referred to as the stochastization energy, or limit, E_{stoch}, which can be considered the lower QC boundary. When the vibrational energy store in the resonant mode exceeds E_{stoch}, the energy spreads over the other modes in accordance with the character of the intermode resonances. This process may recur several times until the total vibrational energy of the molecule grows in excess of the dissociation energy D_0 to open up the unimolecular decay channel. In the case of equilibrium energy distribution over all modes, the vibrational energy store in each nonresonant mode of such a highly excited molecule amounts to a mere D_0/S, where $S = 3N - 6$ is the number of vibrational degrees of freedom of an N-atom molecule. This quantity may even be lower than E_{stoch}. This means that the molecule can rise above E_{stoch} only up the vibrational level ladder of the resonant mode.

In the stochasticity region, the shape of the absorption band of the molecule changes radically in the vicinity of the resonant mode frequency. Owing to intermode interaction, a homogeneously broadened absorption band is formed whose width depends on the intramolecular vibrational energy redistribution rate, and the central frequency is shifted towards the long-wavelength region by an amount depending on the average *intermode anharmonicity*. Both the width and shift of the vibrational band increase as the vibrational energy of the molecule becomes higher, but as a rule, the shift does not perceptibly exceed the width. This is due to the fact that far from all vibrational modes of the molecule contribute to the intermode anharmonicity (the model of the active and passive modes of the vibrational reservoir [14]). It is exactly this relationship between the width and shift of the fundamental IR absorption band of a polyatomic

molecule that ensures the possibility of its absorbing a large number of IR photons.

The IR MP excitation model presented in Fig. 4 elaborates the initial, simple model in two aspects. First, it introduces the notion of the stochastization energy, E_{stoch}, which is determined by the density of the intermode Fermi resonances and the intermode anharmonicity. And secondly, it introduces the notion of a homogeneously broadened fundamental absorption band at the molecular vibrational energy $E_V > E_{stoch}$, the band width 2γ and the shift δ increasing with the growing vibrational energy of the molecule. Recent theoretical works have made it possible to develop models for calculating E_{stoch}, $\gamma(E_V)$, and $\delta(E_V)$ and relating them to the parameters of anharmonicity, which is an essential step in the quantitative description of the IR MP excitation of polyatomic molecules [15, 16].

A highly vibrationally-excited molecule can undergo, first, various monomolecular reactions (such as dissociation, fragmentation and isomerization), and

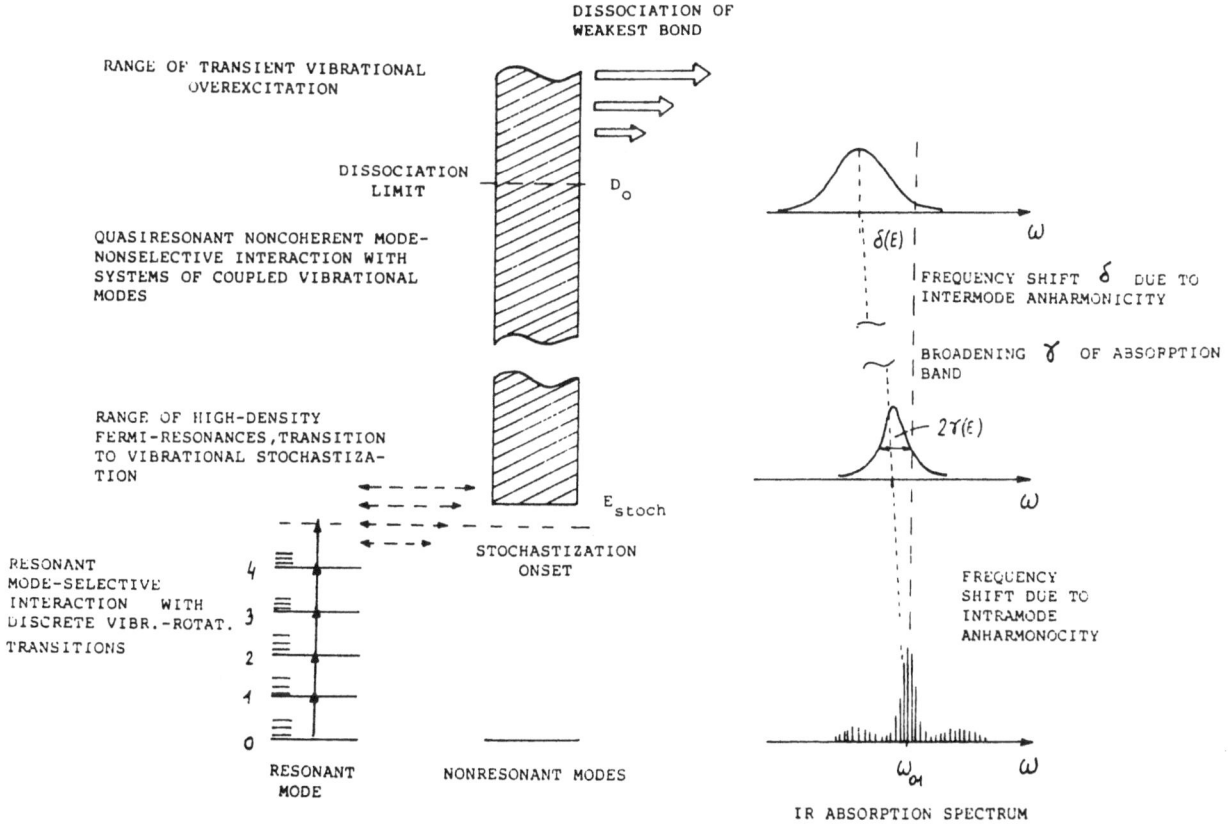

Fig. 4. A more detailed model of photodissociation of a polyatomic molecule by an intense IR field: left – scheme of vibrational energy acquisition by the molecule in the region of "mode-selective" and "mode-nonselective" excitation; right – evolution of the fundamental IR absorption band spectrum with the increasing vibrational energy of the molecule. Even at the dissociation limit the molecule is capable of absorbing, in a quasi-resonant fashion, IR radiation at the laser frequency ω_L tuned to the long-wavelength wing of the fundamental absorption band

then the decomposition products of a polyatomic molecule, such as active radicals, can themselves participate in subsequent reactions.

Below are listed some features of the processes of IR MP excitation and dissociation of polyatomic molecules which are most important as far as industrial applications are concerned.

Universality

IR MP processes can be observed in polyatomic molecules of any complexity and symmetry if they have more than four or five atoms and have IR absorption bands in the spectral region of lasing of high-power IR lasers. Most experiments on MP excitation and MP dissociation have been carried out with TEA CO_2 lasers ($\lambda = 9$–$10\ \mu m$). The universality of IR MP processes allows the dissociation of many various molecules and the production of a wide variety of free radicals and provides a potential basis for a variety of processes for IR laser radical chemical synthesis.

Intermolecular Selectivity

Intense IR radiation is able to deposit the energy essential for the dissociation of wanted molecules in a mixture without depositing such high energy into the other molecules in the gas mixture. This is caused first, by the strong difference in MP absorption spectra in different molecules and, second, by the possibility of realizing collisionless excitation during an IR pulse when there is no significant collisional energy transfer from the excited molecule to the other molecules in the mixture. This most important feature of IR MP processes is successfully used in laser isotope separation [17]. From the standpoint of IR laser radical synthesis, intermolecular selectivity is also very important. It leads a strong intermolecular nonequilibrium in the reacting molecular mixture and a high degree of directivity of laser radical reactions [18].

Vibrational Energy Stochastization

A high-power IR field involves all or, at least, many modes of polyatomic molecule in the process of vibrational excitation. This conclusion follows from numerous direct and indirect experiments concerned with intramolecular vibrational energy redistribution. As the MP excitation is performed with IR radiation pulses 10^{-6}–10^{-8} s long with $W_{exc} \simeq 10^7$–$10^9\ s^{-1}$, the vibrational motion of polyatomic molecules excited to the dissociation limit is stochastized and can be described by a statistical method [19]. Indeed, in all reliable experiments of dynamics of molecular IR dissociation it was impossible to observe any essential

difference from the predicted statistical theory of monomolecular decay, RRKM. This, among other things, means that the dissociation of MP-excited molecules generally occurs due to a break in the weakest bond.

Difference in Vibrational and Translational Temperatures

The laser radiation during the process of collisionless IR MP excitation increases only the vibrational energy of the molecule, its translational energy remaining almost constant. This situation differs from the case of purely thermal initiation in which all the degrees of freedom of the molecule (vibrational, translational, rotational) have the same high temperature.

Each of these properties provides new possibilities for chemists.

3. Photochemistry in Excited Electronic States Induced by IR + UV Excitation

As a rule, the electronic absorption bands of molecules are broad and structureless under standard experimental conditions. Therefore they are useless for selective excitation, since they usually do not provide the required degree of spectral selectivity. If the excited electronic state is to be used, the most universal and efficient way to achieve the required selectivity is to selectively excite the vibrational state which subsequently leads to the transfer of molecules into the necessary excited electronic state. The required selectivity is acquired in the first step of the photoexcitation while the necessary type of photochemical process (dissociation, isomerization or ionization) is determined in the second step (Fig. 2c).

The first experiments on the two-step IR-UV selective photodissociation and isotope separation of $^{14}NH_3$ and $^{15}NH_3$ were accomplished using this method in 1972 [20]. Powerful IR lasers allow the excitation of the high vibrational levels of polyatomic molecules.

The excitation of several vibrational levels during MP absorption of intense IR radiation results in a much larger "red" shift of the UV absorption band than that observed after the absorption of one IR photon. This method of selective MP IR excitation and subsequent UV dissociation has been realized for the molecules $^{12}CF_3I$ and $^{13}CF_3I$ [21]. In this case the pulse frequency of the CO_2 laser was tuned to the $\bar{V}_1$-absorption band at 9.6 μm. The $\bar{V}_3$ mode corresponds to the C–I bond and is evident in the UV absorption spectrum. As a result of the redistribution of absorbed vibrational energy among all of the modes

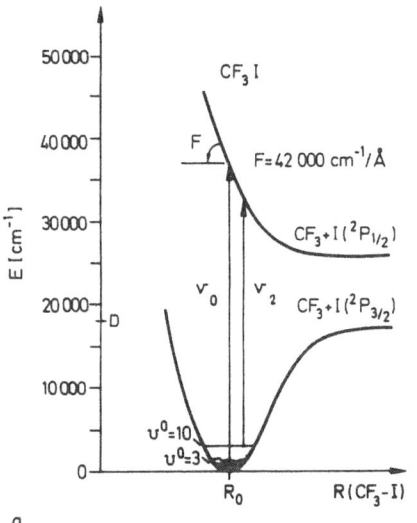

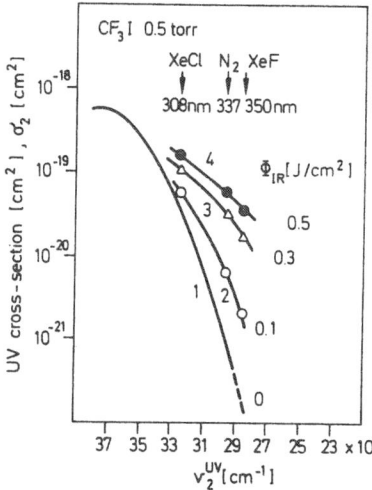

Fig. 5. (a) Potential surface of CF_3I along the C–I bond; (b) Shape of the long-wavelength wing of the UV absorption band of CF_3I at varying energy fluxes of the IR pulse [21]

it also becomes excited. The maximum of the absorption spectrum is at $\lambda_m = 270$ nm. As a result of the UV photon absorption of this band, almost 100% of the molecules dissociate into CF_3 and atomic iodine in the excited state $^2P_{1/2}$ (Fig. 5a). The red shift of the electronic absorption band of this molecule under MP excitation using a 9.6 μm CO_2 laser pulse has also been observed clearly. Figure 5b illustrates the dependence of the cross-section at the long-wavelength edge for different fluxes of IR radiation as well as the absorption cross-section of unexcited molecules [21].

As far as selective photochemistry is concerned, the method of successive IR-UV excitation is rather promising. The application of this method to the separation of the isotopes of heavy elements seems very auspicious [22]. In this case it is necessary to use an IR field of moderate intensity for the preliminarily selective excitation while the dissociation is brought about by UV lasers.

4. Multiple Photon Electron Photochemistry

As in the case of powerful IR irradiation, intense visible or UV radiation induces multiphoton excitation of polyatomic molecules but through the electron intermediate resonant levels. To deposit energy of about 10 eV, for example, a molecule must absorb just 2 or 3 UV photons. Multiphoton excitation of molecules induces various photochemical processes: photofragmentation and photoionization via high-lying electron states, production of active radicals which react with the environment, with the molecules of the solution, etc.

Most photochemical reactions of molecules in solutions are based on exciting the singlet state of the molecules and their subsequent transition to a long-

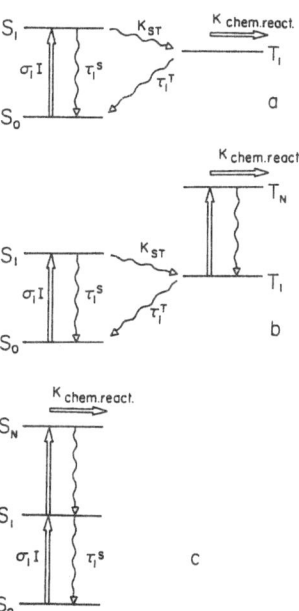

Fig. 6a–c. Excitation schemes of electron states of polyatomic molecules and the corresponding channels of photochemical reaction: (a) single-photon excitation of S_1 singlet state with subsequent transition to the T_1 triplet state; (b) two-step excitation via the intermediate triplet state T_1 under the action of short laser pulses of moderate power; (c) two-step excitation via an intermediate singlet state under the action of high-power ultrashort pulses

lived triplet state (Fig. 6a). The molecules in a long-lived triplet state can absorb another photon and be excited to a high-lying triplet state (Fig. 6b). Such successive absorption of two photons can be achieved with light intensity of $I_{S-T} \simeq \hbar\omega/\sigma_S\tau_1^T\varphi_{isc}$, where σ_S is the cross-section of the electron transition $S_0 - S_1$, τ_1^T is the triplet state lifetime and φ_{isc} is the yield of intercombination conversion. At typical values

$\sigma_S \simeq 10^{-17}$ cm^2 and $\tau_1^T \simeq 10^{-4}$–10^{-7} s the intensity is $I_{S-T} = 10^3$–10^6 W/cm^2, which can be easily obtained with the use of UV lasers generating pulses with $\tau_p \simeq 10^{-8}$ s. The energy fluence of such a pulse $\Phi = I_{S-T}\tau_p \gtrsim 10^{-2}$ J/cm^2 is much lower than the damage threshold of a condensed medium by a laser pulse and hence such two-step excitation is quite realizable.

Under the action of a more powerful laser pulse with an intensity $I_{S-S} \simeq \hbar\omega/\sigma_S\tau_1^S$, where τ_1^S is the excited singlet state lifetime, the probability of absorption of another photon at the next singlet transition $S_1 - S_n$ can become greater than the probability of transition to a triplet state (Fig. 6c). Because of a very short lifetime τ_1^S to 10^{-9}–10^{-11} s the required laser pulse intensity will be $I_{s-s} \simeq 10^9$–10^{11} W/cm^2. Such intensities are quite attainable with the use of powerful lasers generating nanosecond and picosecond pulses. However, it is impossible in practice to apply nanosecond pulses to such two-step excitation of high-lying singlet states since the energy fluence of such a pulse $\Phi = \tau_p I_{s-s} \simeq 1$–$10^2$ J/cm^2. This value considerably exceeds the nonlinear damage threshold of a condensed medium by powerful laser radiation. So one should use powerful picosecond pulses in order to realize photochemical processes via high-lying singlet states. Pulses with a duration $\tau_p \gtrsim \tau_S$ are optimal for realizing highly efficient two-step excitation.

Thus, the photochemical reaction channel can be controlled as a function of the duration of the laser pulse of a given fluence $\Phi \gtrsim \Phi_{sat} = \hbar\omega/\sigma_s$.

5. Laser Isotope Separation

The isotopic selectivity of MP dissociation was first shown in experiments with ^{10}BCl$_3$ and ^{11}BCl$_3$ molecules [7]. In this case the isotope shift in the frequency of the 10 μm vibration being excited was 39 cm^{-1}. A great number of subsequent experiments with different molecules showed that the case of the BCl$_3$ molecule was not unique and that the selectivity of MP dissociation was its characteristic property and resulted from molecules showed that the case of the BCl$_3$ molecule cules by IR laser radiation. In experiments [23] on the separation of the uranium isotopes ^{235}U 'and ^{238}U, isotopical selectivity of dissociation was observed when the isotope shift in the excited mode was just 0.7 cm^{-1}.

Laser isotope separation is one of the most important industrial applications of the selective MP dissociation. Therefore, the great majority of experiments dealing with the study of the kinetics of enrichment, the dependence of selectivity on the conditions of excitation, have been carried out with different isotopical

molecules but the results obtained hold true for mixtures of any other species.

MP excitation and the dissociation of a two-component mixture, for example a mixture of two isotopes, have a number of very essential specific features compared with the case of a one-component medium. The study of the isotopical selectivity of MP dissociation has revealed the means of obtaining maximum selectivity of MP excitation and dissociation. These studies make it possible to gain some insight into the very process of MP excitation of molecules including the subsequent evolution of highly excited molecules.

Many experiments on isotopical selectivity of MP molecular dissociation have thus been performed. These experiments have covered many isotopes, from light ones (hydrogen, deuterium, tritium) to heavy ones (osmium, uranium) contained in very different molecules [8,9]. Many of these experiments became the basis of the laser isotope separation methods developing in numerous laboratories of several countries. Table 1 presents some data on developing processes of isotope separation based on IR MP dissociation, including types of initial molecules and achieved isotopical selectivities of MP dissociation. As an example, let us present more detailed data for the case of ^{12}C and ^{13}C isotope separation which is under development in the USSR through the cooperation of several Institutes (Institute of Spectroscopy, Kurchatov Institute of Atomic Energy and Institute of Stable Isotopes).

To develop an economically viable method of isotope separation it is necessary first of all to choose a polyatomic molecule which satisfies many requirements simultaneously: 1) high yield of MP dissociation for a laser pulse energy fluence which is acceptable for optical windows of laser separation cells (less than 2–3 J/cm^2); 2) high isotopical selectivity of MP dissociation for irradiation at CO$_2$ laser wavelengths; 3) low cost of initial molecular compound. In a present study of a large number of polyatomic molecules (CF$_3$I, CF$_3$Br, ...) it was found that the molecule CF$_2$HCl (Freon-22) is optimal. Parameters for a developed process of carbon isotope separation are given in Table 2.

Figure 7 shows the general scheme of laser set-up for carbon separation. The set-up includes: 1) pulsed periodical CO$_2$ laser with average power up to 5 kW and wavelength in the region of 10 μm which is tunable by a diffraction grating; 2) laser separation cell with fast flow (up to 500 l/s) of gas mixture and with a length of about 5 m in which laser irradiation of ^{13}C-containing polyatomic CF$_2$HCl takes place and 3) chemical separation column in which the separation of dissociation products and initial molecules takes place.

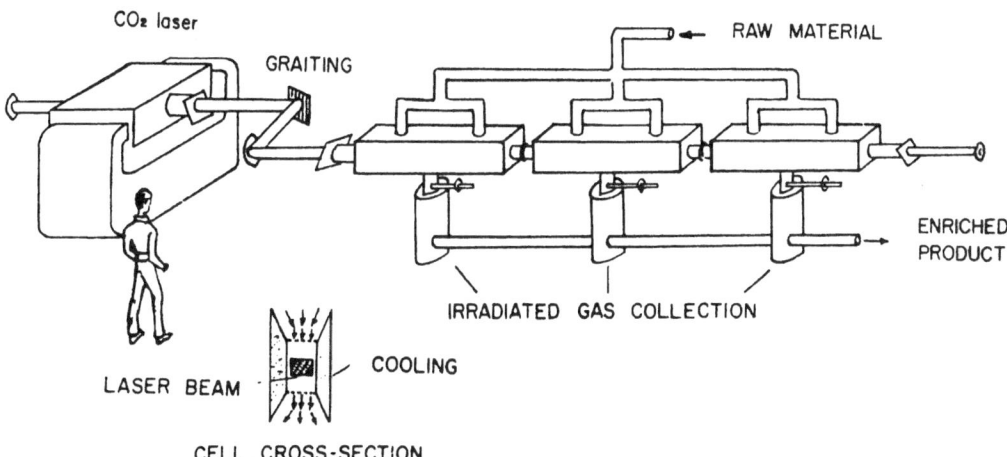

LASER SEPARATION SET-UP

Fig. 7. General view of laser isotope separation set-up for the study of scaling of carbon isotope separation process based on IR MP isotopically-selective photodissociation of molecules using radiation from a high-power CO_2 laser with high rate repetition of laser pulses

Table 1. Developing processes of isotope separation based on IR MP dissociation

Isotope	Content [%]	Mole-cule	Selec-tivity	Country
D/H	$[D] = 1.5 \times 10^{-2}$	CF_3D CF_2DCl	$\gtrless 10^4$	USA Canada
T/D	$[T] = 6 \times 10^{-5}$	CF_3T CCl_3T	$\gtrless 10^4$	Japan USA
$^{10}B/^{11}B$	$[^{10}B] = 19$	BCl_3 $HClCCHBCl_2$	80	USSR USA
$^{13}C/^{12}C$	$[^{13}C] = 1.1$	CF_2HCl	6×10^3	USSR Canada
$^{34}S/^{36}S$	$[^{34}S] = 4.2$	SF_6	35	USSR
$^{235}U/^{238}U$	$[^{235}U] = 0.7$	UF_6	2.8	USA

The productivity of the laser separation module is about 100 kg per year of ^{13}C with enrichment up to 80% for 1 kW average power of CO_2 laser.

For reliable operation of the laser separation module it is very important to reduce the requirements on the energy of the laser pulse performing the efficient IR MP dissociation. In this case, first of all we can use the nonfocused beam of the CO_2 laser and irradiate the molecular gas mixture in a long separation cell. Secondly, the probability of laser damage to the optical windows of the separation cell can be diminished. The effective method of reduction of the required energy fluence for IR MP dissociation is using two-frequency irradiation where the first laser pulse performs the isotopically-selective excitation of molecules on a few low-lying vibrational transitions and the second laser pulse excites the molecules in the vibrational quasicontinuum up to the dissociation limit [24]. It is clear from Fig. 4 that the two IR frequency-separated laser pulses might be in much closer resonance with changeable IR absorption band than the single-frequency laser pulse. Two-frequency irradiation of the molecule CF_2HCl improves the parameters of carbon isotope separation significantly (Table 2). Multifrequency irradiation of this molecule gives much better results [25]. In this case very high selectivity of MP IR dissociation (up to 10^4) and high yield of dissociation can be achieved simultaneously. Figure 8 presents the dependences of dissociation yield of the $^{13}CF_2HCl$ molecule on the energy fluence of the CO_2 laser pulse in cases of

Table 2. Parameters for laser process carbon isotope separation

1. Desired	– ^{13}C ($x_0 = 1.1\%$)
2. Method	– MP IR dissociation single frequency field[a]
3. Laser	– pulsed periodical CO_2 laser
4. Starting material	– Freon-22 (CF_2HCl)
5. Enrichment	– 80%[a]
6. Productivity	– 100 kg/year per 1 kW laser power
7. Energy consumption	– 2–4 keV/atom of ^{13}C
8. Cost	– 10 USA \$/g

[a] In the case of two-frequency IR dissociation the enrichment can be higher (up to 98%)

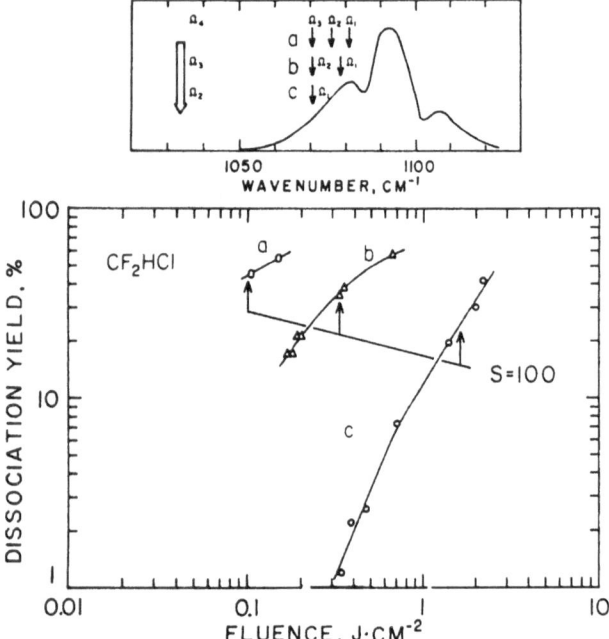

Fig. 8. Dependences of the dissociation yield for the $^{13}CF_2HCl$ molecule on the fluence of the first laser pulse in the cases of (a) double-, (b) triple-, and (c) quadruple-frequency excitation. For all cases, $T = 236$ K, $P(CF_2HCl) = 2.5$ Torr, $P_{buf}(Xe) = 20$ Torr. The arrows indicate the values of yield obtained at a selectivity of 10^2 in the case of double-, triple-, and quadruple-frequency MP dissociation. Positions of laser frequencies for all cases are indicated on the linear IR absorption spectrum of CF_2HCl (upper curve) [25]

2-frequency, 3-frequency and 4-frequency irradiation. Points of required degree of selectivity (10^2) of MP IR dissociation are marked.

6. IR Laser-Induced Chemical Radical Synthesis

The first experiments on IR laser radical chemical synthesis were performed many years ago [7]. As far

back as in [7] in which the isotopical selectivity of IR MP molecular dissociation was first demonstrated, the B_2O_3 molecule was synthesized by the reaction of oxygen with the IR MP dissociation products of the BCl_3 molecule. Later, Clark et al. [26] demonstrated the possibility of radical-radical chemical synthesis of the SF_5NF_2 molecule by acting on a mixture of S_2F_{10} and N_2F_4 molecules with a powerful IR laser pulse. This pulse, with a duration of $\tau_p = 10^{-7}$ s, dissociated both of the initial molecules. The reaction scheme was

$$S_2F_{10} + n\hbar\omega_{IR} \rightarrow 2\,SF_5^{\cdot},$$
$$N_2F_4 + m\hbar\omega_{IR} \rightarrow 2\,NF_2^{\cdot}, \qquad (4)$$
$$SF_5^{\cdot} + NF_2^{\cdot} + M \rightarrow SF_5NF_2 + M,$$

where M is the third body (a molecule, or a cell wall).

Another example of highly effective IR laser radical synthesis has been given in [27], where the isotopically selective dissociation of the CF_3I molecules in the presence of scavengers was investigated. In this work, in particular, the authors demonstrated the possibility of highly effective (practically 100%) photochemical conversion of CF_3Br molecules mixed with I_2 (at a pressure of $I_2 \gtrsim 10$ Torr) to the CF_3I molecule. The photochemical reaction in this case proceeds according to the scheme:

$$CF_3Br + n\hbar\omega_{IR} \rightarrow CF_3^{\cdot} + Br^{\cdot},$$
$$CF_3^{\cdot} + I_2 \rightarrow CF_3I + I^{\cdot}. \qquad (5)$$

Besides CF_3Br and CF_3I, there were no other fluorinated organic compounds in the final products, i.e. the process has a high degree of directivity. Apart from a high conversion level and high degree of directivity, the example under consideration is noteworthy because it demonstrates the possibility of synthesis of a compound (CF_3I) which is thermally less stable than the initial one (CF_3Br) (the dissociation energy of CF_3I is D(C–I) = 56 kcal/mol and that of CF_3Br is D(C–Br) = 68 kcal/mol).

To ensure an optimal process of laser radical chemical synthesis, it is necessary to provide conditions such that during the IR MP dissociation of the initial molecule only radicals of the wanted sort are formed and the radiation energy is not expanded for the formation of undesired secondary radicals. In the above-mentioned cases during the IR MP dissociation of such relatively simple molecules as BCl_3, CF_3I, and CF_3Br, it really takes place, and it is due to two factors. First, the free radicals formed by the IR MP dissociation of these molecules do not possess strong absorption bands coinciding with the excited bands of the initial molecules. Second, the resulting radicals consist of few atoms, which also reduces their further excitation and dissociation.

However, as the experiments on IR MP dissociation of a large number of molecules [9] show, IR MP excitation often produces several types of radicals. First of all, this is possible only when the radical resulting from the primary act of dissociation can be subjected to further decay with smaller fragments formed. This, in turn, is possible for two reasons: first, due to thermalization of primary radicals during their collisions with highly excited initial molecules and, secondly, due to further absorption of IR radiation by radicals. Also, in some cases a molecule has several dissociation channels with nearly equal activation energies. Corresponding dissociation rates become comparable, leading to the competition of different parallel channels of dissociation.

These processes are undesirable in laser radical synthesis since they cause the desirable product yield to decrease. In this connection let us consider the IR laser-induced synthesis of the $(CF_3)_3CI$ molecule [28]. This case gives an illustrative example of the possibility of *selective* molecular dissociation as applied to chemical synthesis.

The $(CF_3)_3CI$ molecule was synthesized by irradiating a mixture of $(CF_3)_3CBr$ and I_2 molecules with TEA CO_2 laser pulses. It is relatively difficult to carry out ordinary thermal synthesis of the $(CF_3)_3CI$ molecule because it calls for the use of highly toxic substances and, also, has a quite low yield of the desirable product ($\simeq 1\%$). In the presence of I_2 IR MP dissociation of $(CF_3)_3CBr$ generates both the desired radical $(CF_3)_3C\cdot$, which reacts to form the desired product $(CF_3)_3CI$, and the undesired secondary radical $CF_3\cdot$, which reacts to form the secondary product CF_3I. Figure 9 presents the dependence of the dissociation yield Y_{diss} and the relative yield of desired product Y_{synth}/Y_{diss} (i.e., the fraction of dissociation $(CF_3)_3CBr$ molecules which have been spent on the formation of the desired product $(CF_3)_3CI$) as a function of the laser energy fluence Φ. The value Y_{synth}/Y_{diss} is a convenient parameter for the degree of directivity of the process for the formation of the desired product $(CF_3)_3CI$. The value Y_{diss} grows quickly and the value Y_{synth}/Y_{diss} drops as the radiation energy fluence increases.

To interpret the presented experimental dependences, let us consider the reaction scheme. The primary act of IR MP dissociation of $(CF_3)_3CBr$ molecules is the detachment of the Br atom

$$(CF_3)_3CBr + n_1 \hbar \omega_{IR} \rightarrow (CF_3)_3C\cdot + Br\cdot . \qquad (6)$$

The synthesis of $(CF_3)_3CI$ occurs in the reaction

$$(CF_3)_3C\cdot + I_2 \rightarrow (CF_3)_3CI + I\cdot . \qquad (7)$$

The undesired secondary radical CF_3 can be formed, in principle, in the following reactions: the parallel

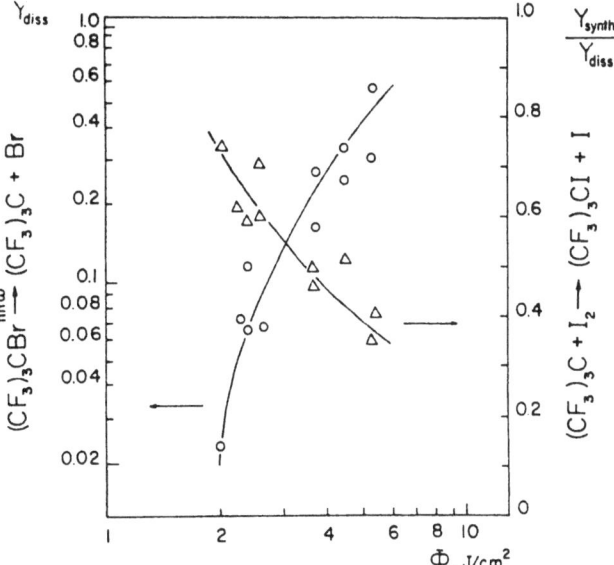

Fig. 9. Dependences of yield of MP IR dissociation of $(CF_3)_3CBr$ Y_{diss} (curve *1*) and relative yield of desired product $(CF_3)_3CI$ Y_{synth}/Y_{diss} (curve *2*) as a functions of CO_2 laser energy fluence ϕ. Pressures: $p((CF_3)_3CBr) = 0.9$ Torr, $P(I_2) = 12$ Torr [28]

channel

$$(CF_3)_3CBr + n_2 \hbar \omega_{IR} \rightarrow (CF_3)_2 \cdot CBr\cdot + CF_3\cdot \qquad (8)$$

and the successive channel

$$(CF_3)_3C\cdot + n_3 \hbar \omega_{IR} \rightarrow (CF_3)_2C\cdot + CF_3\cdot . \qquad (9)$$

The successive channel (9) perhaps plays the leading role in the formation of the secondary product [28]. The small contribution of the parallel channel (8) is also supported by the estimation of decay rates according to the RRKM theory. The probability of detachment of the Br atom from

$$(CF_3)_3CBr(D_{C-Br} \simeq 24000\ cm^{-1})$$

exceeds the probability of detachment of the group $CF_3(D_{C-F} \simeq 30000\ cm^{-1})$ by more than one order of magnitude.

IR MP excitation provides the extremely non-equilibrium conditions in the irradiated vessel when the laser energy deposited in different types of degrees of freedom (vibrational not translational) and different molecules in the mixture greatly differs from its equilibrium state. A simple model can explain well the reaction product ratios in this nonequilibrium state. For the sake of simplicity let us consider the reaction of successive decay of a polyatomic molecule ABC proceeding according to the scheme

$$ABC \xrightarrow{k_1} AB + C, \qquad (10)$$

$$AB \xrightarrow{k_2} A + B . \qquad (11)$$

The products of both reactions can be polyatomic and can, generally speaking, go on reacting. However, it is assumed that only reactions (10 and 11) are essential in determining the final concentrations of the products. Molecules AB, A, B or C can also be radicals. In this case it is necessary to assume that the reaction proceeds in the atmosphere of a radical scavenger so that the stable product yield is merely proportional to the nascent concentration of the corresponding radical.

To explain the possibility of a high degree of laser-induced reaction directivity we are going to consider the case when the thermal rate of reaction (11) is high compared with the reaction (10) rate

$$k_2 \gg k_1, \tag{12}$$

that is, the intermediate product AB is thermally unstable. The production of AB is assumed to be the purpose of the chemical synthesis. It is clear that at thermal initiation, reactions (10–11) are not suitable for this purpose since, by virtue of inequality (12), AB decays faster that it is formed. But, if the same reactions are realized under collisionless conditions with IR MP excitation, it is possible to obtain a yield of AB approximating 100%. This is due to the fact that AB molecules are formed cold since their average energy is lower than that of ABC by the value of the energy of the AB–C bond D(AB–C).

Let us determine the desired product, AB, yield

$$Y_{synth} = [AB]/[ABC]_0 \tag{13}$$

and the full consumption, Y_{diss}, of initial substance ABC (i.e., to total yield of both reaction channels)

$$Y_{diss} = 1 - \frac{[ABC]}{[ABC]_0} = \frac{[AB]+[A]}{[ABC]_0}. \tag{14}$$

The connection between these parameters for both types (thermal and IR MP) of excitation is given on Fig. 10 [18, 28].

Calculated dependences of Y_{synth}/Y_{diss} on Y_{diss} show a significant difference between IR MP excitation (curve 2) and thermal initiation (curve 1). Indeed, when the conversion efficiency of the initial substance ABC equals 50% at thermal initiation, the desirable product AB is hardly produced (its yield becomes $(Y_{synth}/Y_{diss})_{1/2} = 3 \times 10^{-3}$). At the same time, in the laser-induced reaction it is practical to convert ABC to the desired product AB almost without the loss of substance ($Y_{synth}/Y_{diss} = 0.998$).

The reaction considered above proceeds under conditions far from thermal equilibrium. The average energy of the molecule ensemble AB formed at thermal dissociation turns out to be much higher than in the case of laser dissociation. Indeed, even though the

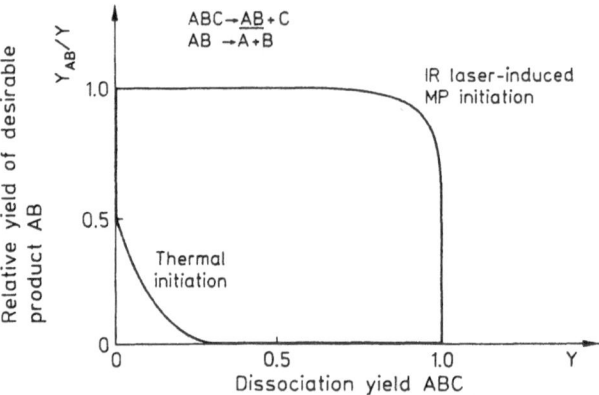

Fig. 10. Dependences of the relative yield of radical AB Y_{synth}/Y_{diss} on the total yield Y_{diss} of dissociation of the molecule ABC in reactions (10, 11) of successive dissociation. Curve 1: thermal initiation. Curve 2: IR MP initiation of reactions [28]

molecules AB are formed directly at the dissociation instant cold, under thermal reaction conditions in collisions they quickly acquire an additional energy, which they lack, to reach the equilibrium energy. This comes about in times shorter than the decay time, and after the dissociation of ABC the product AB is ready, too, for further dissociation. On the other hand, in the case of collisionless laser-induced dissociation the molecules AB remain cold.

7. Picosecond UV Laser-Induced Chemical Reactions

At a high light intensity, i.e. during the excitation of molecules by ultrashort laser pulses, the absorption of even two UV photons and a transition of the molecules to a highly excited electronic state becomes possible. The total energy of the two UV photons is $2\hbar\omega_{UV}^- \gtrsim 8$–$10$ eV, i.e. it corresponds to the energy of excitation by vacuum UV radiation, which does not penetrate into the solvent. This method of multiphoton excitation of molecules thus provides the experimentalist with some new capabilities, which have been demonstrated in several studies.

First, photochemical transformations from highly excited electronic states become possible. These transformations have been observed for molecules of nucleic acids in water [29, 30]. This action on molecules results in photodamage of a new type. It is not observed during illumination with ordinary UV light, which excites only low-lying, primarily triplet states. The basic elementary mechanism for photochemical transformations for nucleic acids was recently identified: the generation of the OH radical in water as a result of the transfer of molecular excitation energy to the solvent (Fig. 11). The mechanism opens up the possibility of developing methods of laser phototherapy (to supple-

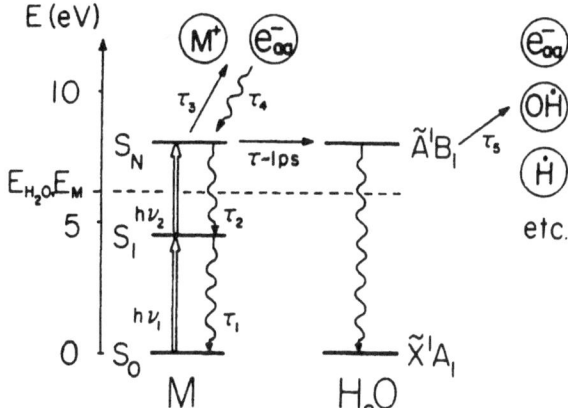

Fig. 11. Two-step UV excitation of high-lying singlet states S_n of biomolecules in water solution. The following primary processes take place: 1) ionization and germinate recombination of produced electron-cation pairs; 2) transfer of electron energy to solvent (water); 3) relaxation of excitation. The main contribution to the formation of final products of photodecomposition of solvated molecules is made by chemical reactions with OH radicals [31]

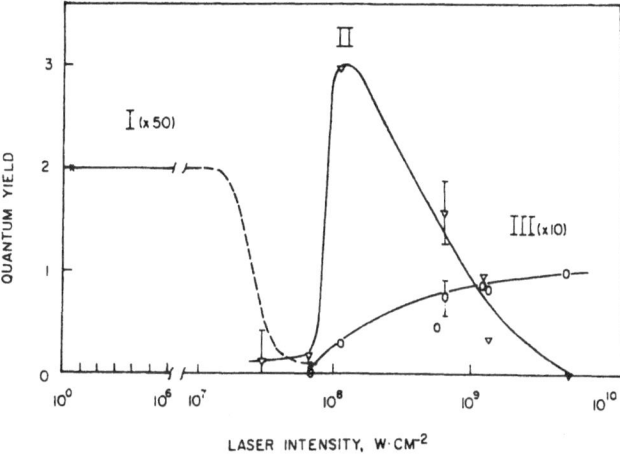

Fig. 12. Dependence of the quantum yield of photoproducts formed from irradiating maleic acid in aqueous solution on UV radiation intensity: *I* formation of fumaric acid due to cis-transisomerization of maleic acid at low intensity; *II* formation of maleic acid dimers due to radical dimerization under irradiation with nanosecond pulses of moderate intensity; *III* formation of malic acid under irradiation with picosecond pulses of high intensity [33]

Second, the possibility of controlling the pathways for photochemical reactions by varying the intensity of the laser light (Sect. 4) has been observed in the experiments on the photochemical synthesis of amino-acids and hydroxy acids through the application of intense pulses of UV laser light of picosecond and nanosecond duration [32, 33]. In [32], for example, it has been found that, as the aqueous solution of ammonium salt of maleic acid is irradiated by powerful UV (265 nm) picosecond radiation, ammonia adds to the double bond of acid $-C=C-$ yielding to the corresponding α-amino-acid

$$\begin{array}{ccccccc} O & H & H & O & & O & H & H & O \\ \| & | & | & \| & \xrightarrow{2\hbar\omega} & \| & | & | & \| \\ C-C=C-C & & & & & C-C-C-C \\ | & & | & & & | & | & | & | \\ H_4NO & & ONH_4 & & & OH & H & NH_2 & OH \end{array} \tag{15}$$

It is essential that the yield of such a photochemical reaction stimulated by two-step excitation of molecules via an intermediate singlet state may be as high as 100%.

In the experiments on nonlinear UV photochemistry of maleic acid in aqueous solution [33] it has been shown that as the UV laser radiation intensity increases, the channels of photochemical reactions change. Comprehensive studies on the photochemistry of maleic acid under low-intensity UV light have been carried out [34]. It is known that in this case in aqueous solution the isomerization of maleic acid through the excitation of a triplet state comes about (region I, Fig. 12):

ment the existing methods of radio-therapy which use ionizing radiation) through the excitation of electronic states of biological molecules which lie above the ionization limit of water molecules [31].

maleic acid   HOOC   COOH   H   COOH   fumaric acid
(cis-isomer)   \\   /   \\   /
                    C=C $\rightleftharpoons$ C=C
                   /   \\   /   \\
                  H   H   HOOC   H   (trans-isomer)

(16)

Under the action of nanosecond UV pulses (265 nm) photodimerization was observed which occurs due to two-step excitation via triplet states to the region of intensities from 3×10^7 to 10^9 W/cm^2 (region II, Fig. 12). Dimers are formed due to breaking of the double bond $-C=C-$

$$\begin{array}{c} \overset{H}{\underset{|}{}} \ \overset{H}{\underset{|}{}} \\ HOOC-C{=}C-COOH \\ HOOC-C{=}C-COOH \\ \underset{H}{\overset{|}{}} \ \underset{H}{\overset{|}{}} \end{array} \xrightarrow{2\hbar\omega} \begin{array}{c} \overset{H}{\underset{|}{}} \ \overset{H}{\underset{|}{}} \\ HOOC-C-C-COOH \\ HOOC-C-C-COOH \\ \underset{H}{\overset{|}{}} \ \underset{H}{\overset{|}{}} \end{array} \qquad (17)$$

Under the action of picosecond UV pulses (265 nm) with their intensity ranging from 10^8 to 6×10^9 W/cm^2 a reaction of water addition due to the excitation of singlet molecular states takes place (region III, Fig. 12).

$$\begin{array}{c} HOOC \qquad\quad COOH \\ \diagdown \qquad\quad \diagup \\ C {=\!=} C \\ \diagup \qquad\quad \diagdown \\ H \qquad\qquad H \end{array} \xrightarrow{2\hbar\omega} HOOC-\overset{\overset{\displaystyle H}{|}}{C}-\overset{\overset{\displaystyle H}{|}}{\underset{\underset{\displaystyle OH}{|}}{C}}-COOH. \qquad (18)$$

Thus, by changing only one radiation parameter, intensity, it is possible to stimulate qualitatively different types of photochemical transformations and to selectively produce photochemical products with a high quantum yield.

Photochemical transformations of molecules with multiple $-C=C-$ bonds under the action of high-power UV picosecond laser pulses are worthy of detailed study since, first, olefins are one of the widespread classes of chemical compounds and, secondly, photochemical reactions with double-bond breakage open the way for photochemical synthesis of many compounds which cannot be synthesized by the methods of single-photon laser photochemistry.

8. Conclusions

So, in this short review we have discussed the basic photoexcitation processes and a few potential applications of laser-induced processes in chemistry. The emphasis was on multiple-photon laser-induced processes, including the excitation of high-lying vibrational or electronic states of polyatomic molecules, i.e. on nonlinear photochemical processes. Unique combinations of valuable parameters in laser light sources open interesting possibilities even in the field of nonlinear photochemistry using single-photon excitation. Considerable progress is being made in UV photochemistry with excimer lasers. Let us mention the recent progress in the synthesis of C_2H_3Cl by

excimer laser-induced radical-chain reactions [35]. This process is of considerable industrial importance, being the main route in the production of vinyl chloride monomer feedstock for PVC manufacture. In this case the excimer KrF laser generates the enhanced concentration of free radicals which control the chain reaction of 1,2-dichlorethane (DCE) to vinyl chloride (VC) conversion. The experiments demonstrate clearly the effect of laser-generated free radicals in the DCE to VC conversions used in the technical process.

Another example of progress in UV photochemistry with excimer lasers is a demonstration of photochemical ablation of organic materials. When a nanosecond pulse of UV radiation of excimer lasers (wavelength 193 nm) with a fluence above a threshold value (less than 1 J/cm^2) falls on a organic polymer film, the material at the irradiation site is etched [36]. This process has been called ablative photodecomposition. UV laser ablation of polymers promises to be an exceedingly interesting phenomenon for further study and potential applications in technology [37]. This UV laser-induced photoprocess is also quite efficient for the destruction of such bio-organic materials as tissue, plaque, etc. and has a great potential application in laser surgery, particularly in the treatment of cardiovascular diseases (laser angioplasty). Recent research in the application of lasers to atherosclerosis has shown that the high-energy ultraviolet output of an excimer laser can be successfully used to identify a blocked artery segment in vitro [38]. Study of comparable efficiency and products of atherosclerotic plaque destruction by pulsed laser radiation of various wavelengths has been performed [39]. We can expect significant progress in this type of application of laser photochemistry in medicine in the near future.

References

1. J.C. Calvert, J.N. Pitts, Jr.: *Photochemsitry* (Wiley, New.York 1966)
2. A. Turro: *Molecular Photochemistry* (Benjamin, New York 1965)
3. R.V. Ambartzumian, V.S. Letokhov: Appl. Opt. **11**, 354 (1972)
4. V.S. Letokhov: Science **180**, 451 (1973)
5. V. Merchant, N.R. Isenor, R.S. Hallsworth, M.C. Richardson: Can. J. Phys. **51**, 1281 (1973)
6. R.V. Ambartzumian, N.V. Chekalin, V.S. Doljikov, E.A. Ryabov: Chem. Phys. Lett. **25**, 515 (1974)
7. R.V. Ambartzumian, V.S. Letokhov, E.A. Ryabov, N.V. Chekalin: Pis'ma Zh. Eksp. Teor. Fiz. (Russ.) **20**, 597 (1974)
8. V.S. Letokhov: *Nonlinear Laser Chemistry with Multiple Photon Excitation*, Springer Ser. Chem. Phys. **22** (Springer, Berlin, Heidelberg 1983) p. 417
9. V.N. Bagratashvili, V.S. Letokhov, A.A. Makarov, E.A. Ryabov: *Multiple Photon Infrared Laser Photophysics and Photochemistry* (Harwood, London 1985) p. 512
10. R.V. Ambartzumian, V.S. Letokhov: In *Chem. and Biochem. Applications of Lasers*, Vol. 3, ed. by C.B. Moore (Academic, New York 1977) p. 167
11. N. Bloembergen, E. Yablonovich: Phys. Today **31**, 23 (May 1978)
12. P.A. Schulz, Aa.S. Sudbo, D.L. Krainovich: Annual Rev. Phys. Chem. **30**, 379 (1979)
13. V.B. Chirikov: Phys. Rep. **52**, 263 (1979)
14. V.N. Bagratashvili, M.V. Kuzmin, V.S. Letokhov, A.A. Stuchebrukhov: Chem. Phys. **97**, 13 (1985)
15. A.A. Stuchebrukhov, M.V. Kuzmin, V.N. Bagratashvili, V.S. Letokhov: Chem. Phys. **107**, 429 (1986)
16. M.V. Kuzmin, V.S. Letokhov, A.A. Stuchebrukhov: Zh. Eksp. Teor. Fiz. (Russ.) **90**, 458 (1986)
17. E.P. Velikhov, V.Yu. Baranov, V.S. Letokhov, E.A. Ryabov, A.N. Starostin: *Powerful Pulsed CO_2 Lasers and their Application for Isotope Separation* (Russ.) (Publ. House "Science", Moscow 1983)
18. V.N. Bagratashvili, M.V. Kuzmin, V.S. Letokhov: J. Phys. Chem. **88**, 5780 (1984)
19. I. Oref, B.C. Rabinowitch: Acc. Chem. Res. **12**, 166 (1979)
20. R.V. Ambartzumian, V.S. Letokhov, G.N. Makarov, A.A. Puretzky: Pis'ma Zh. Eksp. Teor. Fiz. (Russ.) **15**, 709 (1972); **17**, 91 (1973) [Sov. Phys.-JETP Lett. **15**, 501 (1973); **17**, 63 (1973)]
21. I.N. Knyazev, Yu.A. Kudriavtzev, V.S. Letokhov, A.S. Sarkisian: Appl. Phys. **17**, 427 (1978); Zh. Eksp. Teor. Fiz. **76**, 1281 (1979) (Russ.)
22. R.J. Jensen, O. Judd, J.A. Sullivan: Los-Alamos Science **3**, 2 (January 1982)
23. J.A. Horsley, D.M. Cox, R.B. Hall, A. Kaldor, E.T. Maas, Jr., E.B. Priestly, G.M. Kramer: J. Chem. Phys. **73**, 3660 (1980)
24. R.V. Ambartzumian, N.P. Furzikov, Yu.A. Gorokhov, V.S. Letokhov, G.N. Makarov, A.A. Puretzky: Opt. Commun. **18**, 517 (1976)
25. A.V. Evseev, V.S. Letokhov, A.A. Puretzky: Appl. Phys. B **36**, 93 (1985)
26. J.H. Clark, K.M. Leary, T.R. Loree, L.B. Harding: In *Advances in Laser Chemistry*, ed. by A.H. Zewail, Springer Ser. Chem. Phys. **3** (Springer, Berlin, Heidelberg 1978) p. 74
27. G.I. Abdushelishvili et al.: Pis'ma Zh. Techn. Fiz. (Russ.) **5**, 849 (1979)
28. V.N. Bagratashvili, M.V. Kuzmin, V.S. Letokhov: Laser Chem. **4**, 139 (1984)
29. P.G. Kryukov, V.S. Letokhov, D.N. Nikogosyan, A.V. Borodavkin, E.I. Budowsky, N.A. Simukova: Chem. Phys. Lett. **61**, 375 (1979)
30. D.N. Nikogosyan, V.S. Letokhov: Riv. Nuovo Cimento **6**, 1 (1983)
31. D.N. Nikogosyan, A.A. Oraevsky, V.S. Letokhov: Lasers Life Sci. **1**, 49 (1986)
32. V.S. Letokhov, Yu.A. Matveets, V.A. Semchishen, E.V. Khoroshilova: Appl. Phys. B **26**, 243 (1981)
33. E.V. Khoroshilova, N.P. Kuzmina, V.S. Letokhov, Yu.A. Matveetz: Appl. Phys. B **31**, 145 (1983)
34. J.N. Pitts, R.L. Letsinger, R.P. Tayler, J.M. Paterson, G. Recktenwals, R.B. Martin: J. Am. Chem. Soc. **81**, 1068 (1959)
35. K. Kleinermans, J. Wolfrum: Laser Chem. **2**, 339 (1983)
36. R. Srivanasan, V. Magne-Banton: Appl. Phys. Lett. **41**, 576 (1982)
37. R. Srivanasan: In: *Laser Processing and Diagnostics*, ed. by D. Bäuerle, Springer Ser. Chem. Phys. **39** (Springer, Berlin, Heidelberg 1984) p. 343
38. J.M. Isner, R.H. Clarke: IEEE J. QE-**20**, 1471 (1981)
39. N.P. Furzikov, T.I. Karu, V.S. Letokhov, A.A. Beljaev, S.E. Ragimov: Lasers Life Sci. **1**, 265 (1987)

Appl. Phys. B 46, 253–260 (1988)

Applied
Physics B

Photo-
physics
and Laser
Chemistry

© Springer-Verlag 1988

Atomic Vapor Laser Isotope Separation*

J. A. Paisner

Laser Isotope Separation, Lawrence Livermore National Laboratory,
P.O. Box 5508, Livermore, CA 94550, USA

Received 19 April 1988/Accepted 25 April 1988

Abstract. Atomic Vapor Laser Isotope Separation (AVLIS) is a general and powerful technique applicable to many elements. A major present application to the enrichment of uranium for lightwater power reactor fuel has been under development at the Lawrence Livermore National Laboratory since 1973. In June 1985, the Department of Energy announced the selection of AVLIS as the technology to meet future U.S. needs for the internationally competitive production of uranium separative work. Major features of the AVLIS process will be discussed with consideration of the process figures of merit.

PACS: 32.80-T

Atomic Vapor Laser Isotope Separation (AVLIS) is a general process for converting a feed stream into a product stream in which a selected set of isotopes has been enriched or depleted [1]. The heart of the process is the selective multistep photoionization of an atomic vapor stream. The components of a generic AVLIS process are shown in Fig. 1. The process hardware is divided into a separator system and a laser system that are, to a great degree, mechanically independent. Atomic vapor is produced in the vaporizer and expands upwards in vacuum. Tunable laser frequencies are generated in a dye laser system that is in turn driven by a pump laser system. Copper lasers serve as the pump lasers in the major systems we have constructed. Both the pump lasers and tunable lasers are configured in master-oscillator/power-amplifier (MOPA) chains. The laser light illuminates the atomic vapor near the surface of an ion extractor. Photoions are drawn to and neutralized at these electrically biased surfaces. The remaining vapor streams through to the roof of the separator. In an enrichment mission the material from the extractor is enriched product; the material from the roof is depleted tails. In a stripping mission, product and tails are reversed. At Livermore, we have investigated the application of this laser isotope separation technology to a number of missions including those for defense programs. Of particular interest are applications with reasonable product demand, some of which are shown in Fig. 2. The dominant mission is clearly the enrichment of uranium for use in civilian light-water reactors. Laser isotope separation technology for uranium enrichment has been under development at Livermore since 1973 in a joint effort with

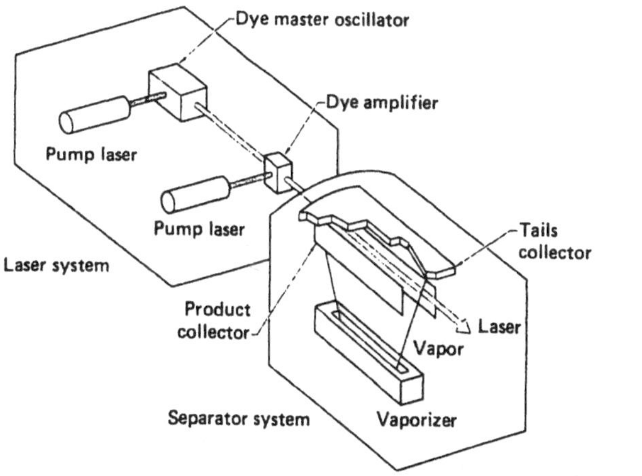

Fig. 1. Schematic of major subsystems employed in the Atomic Vapor Laser Isotope Separation (AVLIS) process at Livermore

---

* This paper is a synopsis of a presentation made at the Royal Swedish Academy of Engineering Sciences IVA Symposium "Laser Technology in Chemistry," Stockholm, Sweden, November 10, 1987

Element	Application	Demand (Mg/y)
Uranium	Low-cost fuel for light water reactors	>1000
Samarium Europium Gadolinium, etc.	Burnable poison for power reactors	>1
Mercury	More efficient fluorescent lamps	>1
Zirconium	Cladding for nuclear fuel elements	>1
Rhodium Palladium Platinum	Precious metal recovery from nuclear waste	>1

Fig. 2. Candidate elements for processing using atomic vapor laser isotope separation

Martin Marietta Energy Systems, Oak Ridge, Tennessee. In June 1985, following an intensive year-long review, the Department of Energy selected AVLIS as the technology to meet the nation's future need for competitive production of enriched uranium.

1. Process Figures of Merit

The figures of merit of an enrichment process can be understood by examining the SWU (Fig. 3). The SWU is a value function with units of mass that depends on the mass flows and assays. The expression comes from cascade theory. It is related to, but not proportional to, the entropy of mixing destroyed by the enrichment process. For natural uranium feed, light-water-reactor fuel product, and typical tails assay, 1 SWU converts 1.24 kg of feed into 0.21 kg of product. Consequently, a world demand of 7000 Mg of product/year corresponds to processing 40,000 Mg of feed/year and producing 3×10^7 SWU. Hence any uranium-enrichment process involves large-scale hardware.

A reasonable target cost for separative work is $60/SWU. This corresponds to a processing cost of about $50/kg of feed or approximately $300/kg of product. Breaking the SWU cost into components as shown in Fig. 4, two components reflect the process engineering costs for the materials-handling and laser systems and two components reflect the process physics, i.e., the laser energy required to process one unit of feed and the separative work obtained from that feed.

To evaluate what is required of the laser system, it is necessary to estimate the MJ of laser energy required to process 1 kg of feed (Fig. 5). Outside of conversion factors, the laser energy is the product of the photon utilization (i.e., the fraction of the required photons that is actually absorbed – accounting for the saturation level necessary to drive the laser/atom kinetics and the light losses in optical elements), the absorbed

$$SWU (kg) = Product (kg) \cdot V(X_p)$$
$$+ Tails (kg) \cdot V(X_T)$$
$$- Feed (kg) \cdot V(X_F)$$
$$V(X) = (2X - 1) \ln \left(\frac{X}{1 - X} \right)$$
$$X = wt \ fraction \ ^{235}U$$

At "standard" conditions
$$X_F = 0.00711$$
$$X_P = 0.032$$
$$X_T = 0.002$$
1 SWU converts 1.24 kg of feed into 0.21 kg of product

Fig. 3. Separative work units

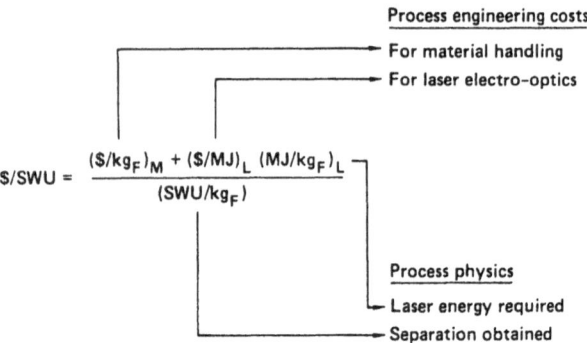

Fig. 4. Elements of separative work cost

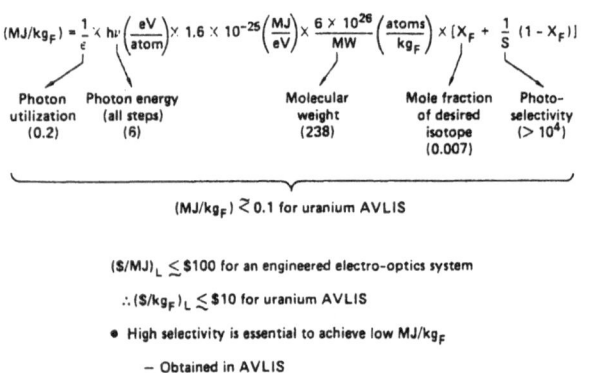

Fig. 5. Laser figures of merit

photon energy needed to process one atom (about 6 eV for any typical photoionization or photodissociation process), and the fraction of the atoms that absorb the light. For a process such as uranium enrichment of natural material, where the isotope of interest is only a very small fraction of the vapor (0.0072 mole fraction), it is essential to have very high photoselectivity. Otherwise, absorption in unwanted isotopes will dominate and require a larger laser system. Putting in values characteristic of the AVLIS process, including the very high process photoselectivity, roughly 100 kJ of laser energy are needed to process 1 kg of uranium feed. The laser cost/kg feed is expected to be on the order of $10.

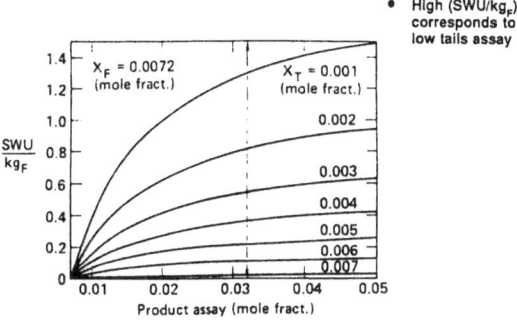

Fig. 6. Separative work performance

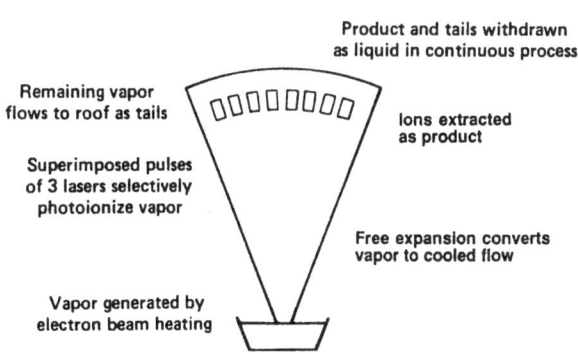

Fig. 7. AVLIS process steps

The other half of the physics performance is the separative work generated per kg of feed. Figure 6, which is a map of separative work performance vs product and tails assay, shows that high SWU production corresponds to low tails assays, not to high product assays. In other words, high selectivity is not essential to achieve high SWU/kg of feed. However, high selectivity is essential to achieve low MJ/kg feed and low \$/SWU. A process that has low feed assay X_F and low selectivity S will have a laser cost $1/(SX_F)$ times higher than a process with high photoselectivity. As an example, a uranium LIS process with a photoselectivity of 2 will require 70 times higher MJ/kg feed and have a concomitantly higher laser-related cost of $> \$ 100/kg$ of feed. The AVLIS process is attractive because it can achieve extremely high process photoselectivity.

The goal for the physics of the AVLIS process is therefore clear; to strip a very high fraction of the ^{235}U atoms out of the feed, leaving few ^{235}U atoms in the tails. All modeling and experiments to date indicate that high stripping efficiency at high photoselectivity and consequently low separative work costs can be obtained using AVLIS.

2. AVLIS Process Steps

The physical steps an atom encounters as it passes through an AVLIS separator are illustrated in Fig. 7. Uranium vapor is generated by an electron beam striking uranium metal held in a cooled crucible. This type of vaporization is commonly used in the metal-plating industry. Nevertheless, uranium is highly refractory (the metal boils at 4100 K) and the technology demands high vaporization rates. The vapor then undergoes adiabatic free expansion into vacuum, in the course of which it transits all of the regimes of aerodynamics: continuum flow near the source, transition flow as the vapor expands, and eventually free molecular flow.

As the vapor passes near the extractor surfaces it is illuminated by three superimposed laser pulses of the wavelengths necessary to drive a three-step photoionization. The approach being used at Livermore is a three-step photoionization process that exploits the large isotope shifts in the electronic spectrum of atomic uranium. These isotope shifts derive primarily from nuclear volume effects. Since the ionization limit is about 6 eV, a three-step process involves lasers operating at about 2 eV or 6000 Å. Consequently, pulsed dye lasers pumped by copper lasers are used in the process. The temporal lengths of the laser pulses are shorter than the radiative lifetimes of the resonant levels in the transition sequence. Very high photoselectivity is attained on each step. Some of the methods of ionization that have been investigated to optimize performance and minimize process laser power requirements are shown in Fig. 8. The net photoselectivity in the process is extremely high as indicated in Fig. 9. The photoionization process can be described as an optical distillation column where work is done only on the isotope or species of interest.

The initial photoplasma contains, to a very good approximation, only ^{235}U ions. If the plasma were left unperturbed, resonant charge exchange between ^{235}U ions and ^{238}U neutrals would in time restore the ion population to natural abundance. This sets a time scale for efficient extraction. The ions go to negatively biased surfaces, and the electrons go to positively biased surfaces or to ground.

In order to obtain the economic advantages of continuous operation, the product stream on the extractors and the tails stream on the roof are collected by liquid flow. This obviously means that the collecting surfaces run at high temperature. In a sense precision laser physics is being performed inside a foundry.

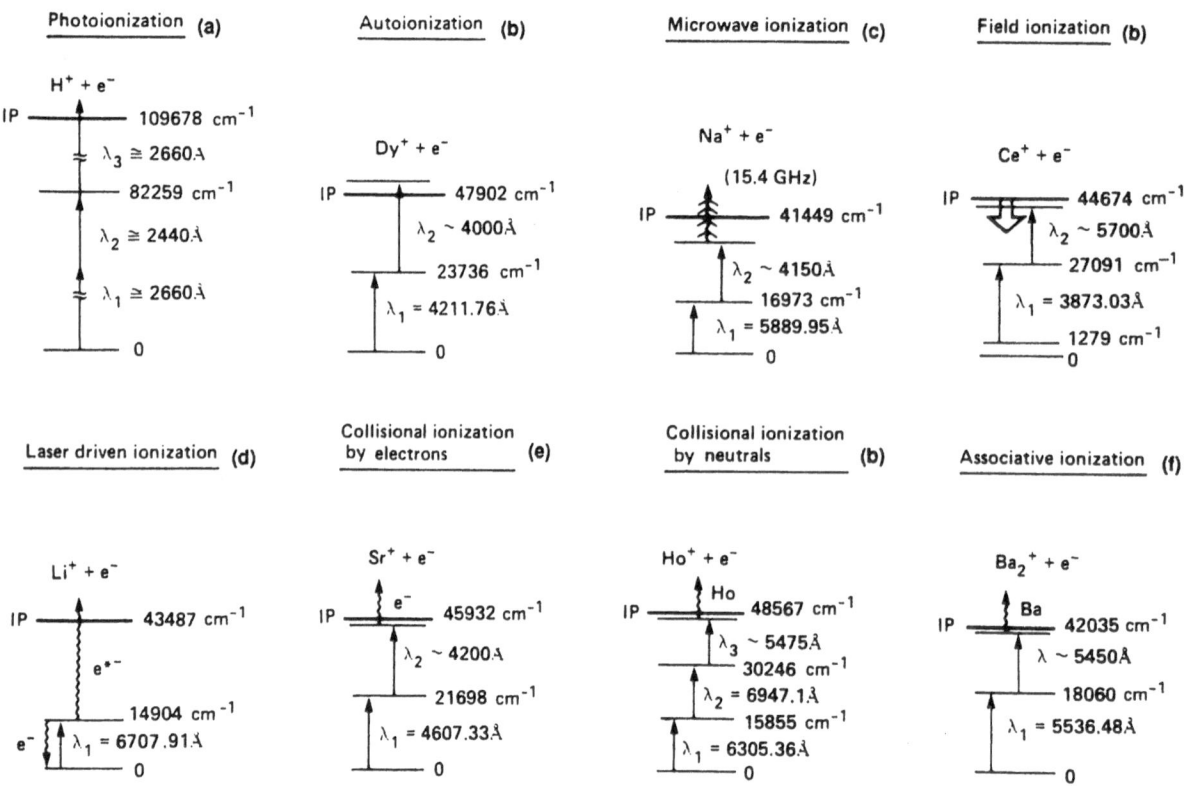

Fig. 8. Methods of atomic ionization after multistep excitation. (a) G.C. Bjorklund, C.P. Ausschnitt, R.R. Freeman, R.H. Storz: Appl. Phys. Lett. **33**, 54 (1978). (b) E.F. Worden, R.W. Solarz, J.A. Paisner, J.G. Conway: J. Opt. Soc. Am. **68**, 52 (1978). (c) P. Pillett, H.B. van Linden van den Heuvell, W.W. Smith, R. Kachru, N.H. Tran, T.F. Gallagher: Phys. Rev. A **30**, 280 (1984). (d) T.J. McIlrath, T.B. Lucatorto: Phys. Rev. Lett. **38**, 1390 (1977). (e) E.F. Worden, J.A. Paisner, J.G. Conway: Opt. Lett. **3**, 156 (1978). (f) R.W. Solarz, E.F. Worden, J.A. Paisner: Opt. Eng. **19**, 85 (1980)

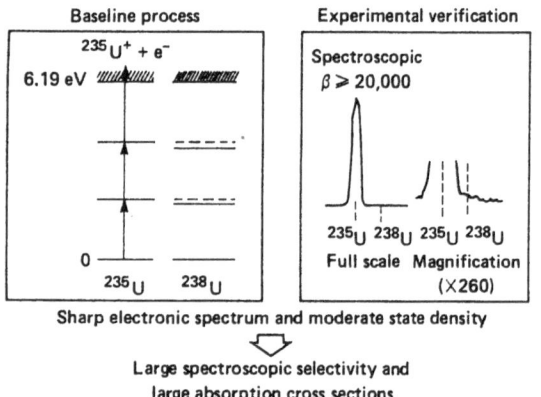

Fig. 9. Spectroscopic selectivity – AVLIS

3. Performance and Cost Modeling

The present generation of systems has been designed and tested after intensive study of the interplay of the physics and engineering that governs the AVLIS process. Each physics area has been modeled, in some cases from first principles, and benchmarked against results obtained in the laboratory or in large-scale enrichment systems described below.

Figure 10 shows an example of the detailed modeling performed in the area of photoionization and propagation physics. The theoretical approach includes a first-principles model based on integrating Schrodinger's equation. This computer model uses as input a complete set of experimentally measured spectroscopic and kinetic parameters corresponding to the relevant transitions of the atom of interest. The code treats the coherent multiple laser/atom interactions and accounts for the evolution of every magnetic sublevel and velocity class within each hyperfine level in the excitation sequence. The code also computes the atomic polarization driven by the light fields and thus facilitates analysis of resonant propagation effects.

These physics models have been incorporated, along with engineering models, in an integrated process model (Fig. 11). The engineering cost models include results from detailed bottom-up costing studies and data obtained from procurements for our demonstration facilities. Also included in the process model are operational parameters based on reliability,

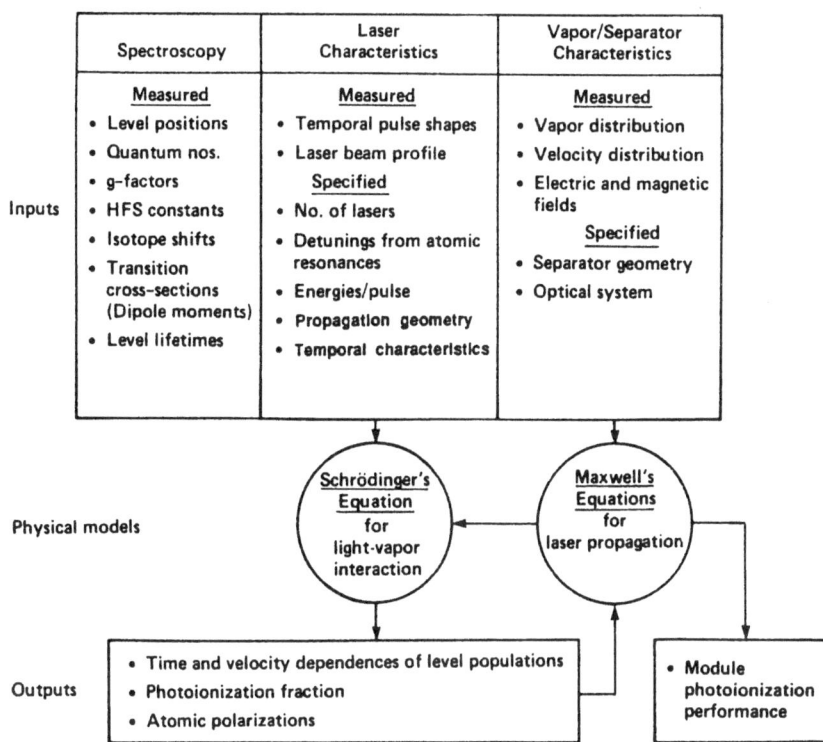

Fig. 10. Modeling of laser-atomic vapor interaction

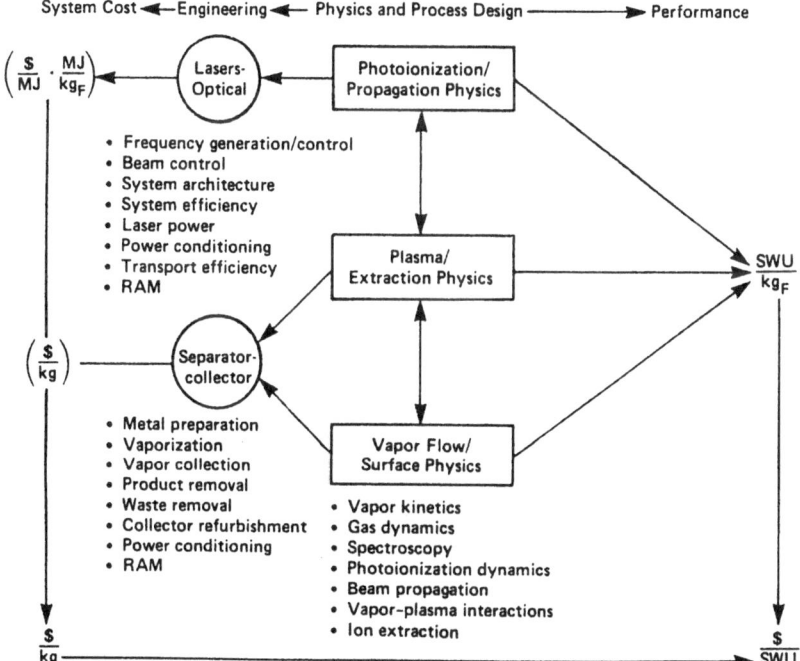

Fig. 11. AVLIS process morphology/structure of integrated process model (IPM)

availability, and maintainability of AVLIS subsystems. Essentially, the integrated process model contains all the fine detail of the AVLIS process and allows an examination of the sensitivity of cost and performance to variations in engineering, design, and physics parameters. There are literally hundreds of parameters that affect the separative work cost of the process. These range from the values of the optical-transition cross sections to the cost of labor. Each one of these parameters has an associated uncertainty and un-

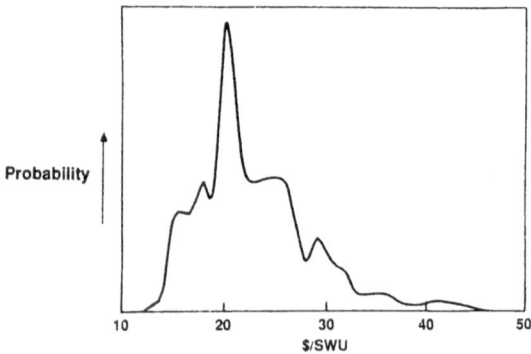

Fig. 12. AVLIS multivariable sensitivity study: distribution of projected separative work cost ~9 million SWU/year (1981 design, 1982 dollars)

certainty distribution. The code is capable of using these distributions in a Monte Carlo calculation of performance and of comparing it to the performance using our base-case design values. Figure 12 illustrates a multivariable sensitivity study histogram for process separative work cost for the 1981 AVLIS engineering design.

4. Development Status

It has taken about 15 years for the AVLIS program to reach its current state of maturity. Process science studies were completed several years ago. In 1985 a full-scale demonstration facility was activated at Livermore. The building, shown in Fig. 13, houses a uranium separator module called the Separator Demonstration Facility. The building also contains a laser called the Laser Demonstration Facility that will provide the laser power for the module. The balance of the building contains instrumentation and control

Fig. 13. AVLIS Full Scale Demonstration Facility consisting of a Laser Demonstration Facility (LDF) and a Separator Demonstration Facility (SDF) at Livermore. The building contains provisions for several copper laser and dye laser corridors and a prototype plant enrichment module.

systems and refurbishment facilities in support of the laser and separator systems.

Figure 14 shows the first completed corridor of copper laser MOPA chains installed in the facility in April 1985. There are 6 MOPA chains containing 30 laser heads (Fig. 15) with a total output capacity of several thousand watts. Figure 16 shows a section of the optical system in the dye laser corridor.

Figure 17 shows the separator module in the Separator Demonstration Facility. The tanks at the

Fig. 14. LDF corridor with copper laser master-oscillator/power amplifiers

Fig. 15. A copper laser operating on a test stand. It operates at approximately 5 kHz and is capable of generating several hundred watts in two colors, 5107 and 5782 Å

Fig. 16. Optical system in dye laser corridor in the Laser Demonstration Facility

Fig. 17. Prototype AVLIS separator module in Separator Demonstration Facility

ends house the module for directing the laser beams through the uranium vapor. The module is essentially plant size and has a projected production rate of over 5×10^5 SWU/year, or 100 Mg of product.

5. Summary

As discussed in this paper, the program at Livermore has focused on laser isotope separation of atomic uranium because of the large demand and high product enrichment price for material used as fuel in commercial light-water nuclear power reactors. In support of that effort, a fundamental approach to the underlying physics was adopted and techniques to obtain the necessary data base were developed. Once developed and deployed for uranium, the AVLIS process can be applied directly to separating many elements economically on an industrial scale. Figure 18 shows the elements whose electronic transitions are accessible using the fundamental and harmonics of the dye system under development at Livermore.

AVLIS has reached industrial scale due to the dedicated efforts of several hundred scientists, engineers and supporting staff. Clearly, the broad range of technologies that comprise AVLIS offer exciting prospects for the future of laser driven processes.

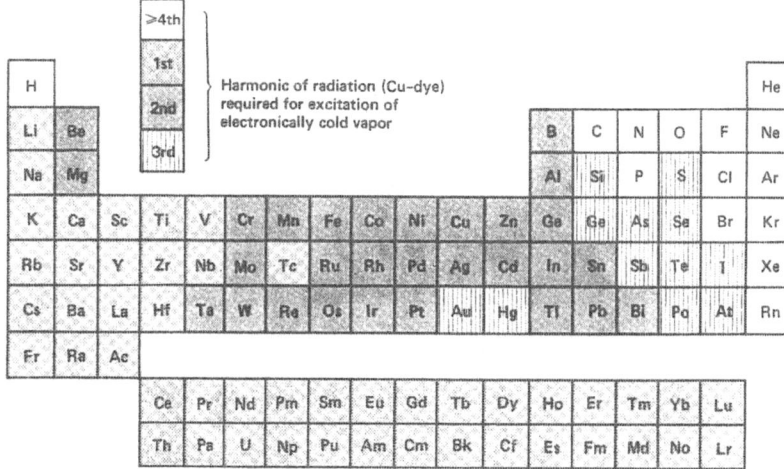

Fig. 18. Elements separable by AVLIS using copper-laser/dye-laser technology

Acknowledgement. Work performed under the auspices of the U.S. Department of Energy by Lawrence Livermore National Laboratory under Contract W-7405-Eng-48.

References

1. Early recognition of the potential of atomic vapor laser isotope separation is generally attributed to Ambartzumian and Letokhov [Ambartzumian 1972] following their experimental work on selective photoionization of rubidium [Ambartzumian, 1971], and Levy and James [Levy, 1973]

R.V. Ambartzumian, V.S. Letokhov: Appl. Opt. **11**, 354 (1972)

R.V. Ambartzumian, V.N. Kalinin, V.S. Letokhov: JETP Lett. **13**, 217 (1971)

R.H. Levy, G.S. James: *Method and Apparatus for the Separation of Isotopes*, U.S. Patent 3,772,519 (1973)

For a more extensive discussion of resonant multistep photoionization, see L.J. Radziemski, R.W. Solarz, J.A. Paisner (eds.) *Laser Spectroscopy and Its Applications* (Dekker, New York 1987), pp. 175

V.S. Letokhov: *Laser Photoionization Spectroscopy* (Academic. Orlando, Florida 1987)

Appl. Phys. B 46, 261–270 (1988)

Chemical Processing with Lasers: Recent Developments

D. Bäuerle

Angewandte Physik, Johannes-Kepler-Universität Linz, A-4040 Linz, Austria

Received 11 March 1988/Accepted 14 March 1988

Abstract. Chemical processing of materials with lasers is a new and interdisciplinary field with many already realized and potential applications in different areas of technology. This overview summarizes some recent developments in this rapidly expanding field.

PACS: 42.60; 68.20; 82.65

The object of laser-induced chemical processing (LCP) of materials is the patterning, coating and physico-chemical modification of solid surfaces by activation of real chemical reactions. Here, the laser-induced activation or enhancement of a reaction may be based on pyrolytical (photothermal) and/or photolytical (photochemical) excitation mechanisms. We shall term a reaction pyrolytic if the thermalization of the laser excitation takes place rapidly in comparison with the reaction, and photolytic if this is not the case, i.e. when the constituents of the reaction are in nonequilibrium states. The laser excitation can take place within the ambient gaseous or liquid medium and/or directly within the surface of the solid material (substrate) to be processed. In many cases the different mechanisms and types of excitation contribute simultaneously to the reaction, although frequently one of them will dominate. There are also many examples where, for example, a reaction is initiated photolytically and proceeds pyrolytically, or vice versa.

Laser light may induce or enhance reactions either heterogeneously in adsorbate-adsorbent systems, at gas-solid or liquid-solid interfaces or within solid surfaces themselves, or homogeneously within surrounding media near solid surfaces. Such reactions may result in materials deposition, surface modification (oxidation, nitridation, reduction, metallization, doping, etc.), materials synthesis, or removal (etching). This is shown schematically in Fig. 1. In the same way as in conventional non-chemical laser processing, chemical processing with lasers can either be performed locally (microchemical processing) or on a more extended scale.

Laser-induced microchemical processing allows for *single-step* direct substrate patterning with lateral dimensions down into the submicrometre range. This can be performed by scanning a focused laser beam across the substrate surface (direct writing), by projecting the laser light via a mechanical mask or by the interference of laser beams. Today, micropatterning is performed by large-area processing techniques, e.g. conventional chemical vapour deposition (CVD) or plasma processing, which are used together with mechanical masking or lithographic methods. Unlike laser microchemical processing, these standard techniques require a large number of production steps. Furthermore, the nonlinearity of laser-induced chemical reactions makes it possible to increase the resolution over that achieved in standard photolithography. In other words, it is possible to produce structures with lateral dimensions smaller than the diffraction-limited diameter of the laser focus.

Large-area chemical processing can be performed either with laser light propagating perpendicular to the substrate surface (normal incidence) and/or, with laser light propagating parallel to the substrate surface. The latter technique allows for extended thin-film processing with or without uniform substrate heating.

In any optical configuration, LCP differs significantly from conventional techniques. Contrary to standard high-temperature techniques such as CVD, LCP makes it possible to restrict or even virtually eliminate heat treatment of the material being processed. Consequently, the laser technique also allows one to process temperature-sensitive materials such as compound semiconductors (GaAs, InP, etc.) and polymer films. Furthermore, LCP avoids the damage to materials – from ion or electron bombardment or from overall vacuum ultraviolet radiation – which is inherent in conventional techniques such as plasma processing and ion- or electron-beam processing. Laser processing is, of course, not limited to planar substrates and allows for three-dimensional fabrication as well. Because of these unique possibilities, LCP is now a rapidly expanding field with many applications in different areas of technology such as in micromechanics, metallurgy, integrated optics, semiconductor device fabrication and chemical technology; in addition, medical applications of laser photochemical techniques are rapidly increasing in number.

In the following we will briefly discuss the various different possibilities of LCP. More detailed descriptions can be found in [1, 2].

1. Deposition of Materials

The various possibilities for laser-induced deposition as schematically shown in Fig. 1 have all been realized. The most detailed investigations, however, were carried out on laser-induced deposition from gas-phase precursors. In the following, we will first briefly discuss the deposition of microstructures with special emphasis on pyrolytic *laser*-induced *chemical vapour deposition* (LCVD), before proceeding to make a few comments on the deposition of thin extended films.

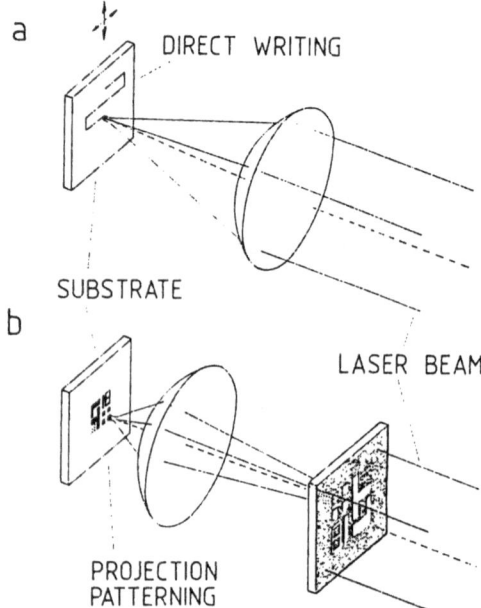

Fig. 2a, b. Schematic showing the optical configuration in direct writing (a) and in projection patterning (b); in the latter case, the optical path is indicated for one feature only

1.1. Microstructures

When a laser beam is focused onto an absorbing substrate (Fig. 2), it induces a local temperature rise and may thereby activate a chemical reaction. If the substrate is immersed, for example, in an atmosphere

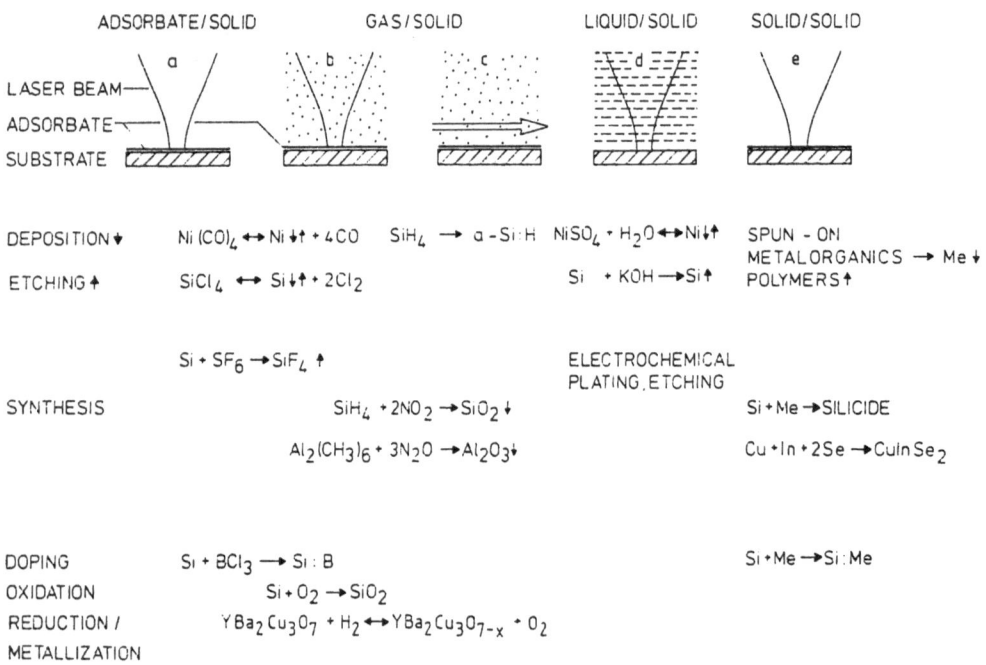

Fig. 1. Examples of laser-induced chemical reactions at or near interfaces. In c), the laser light propagates parallel to the substrate surface. For simplicity, not all reaction products are included in the formulae. The arrows refer to deposition ($\downarrow$) and etching ($\uparrow$). Me stands for metal. $\leftrightarrow$ means that the reaction can be reversed by simply shifting the chemical equilibrium to the other side (adapted from [1])

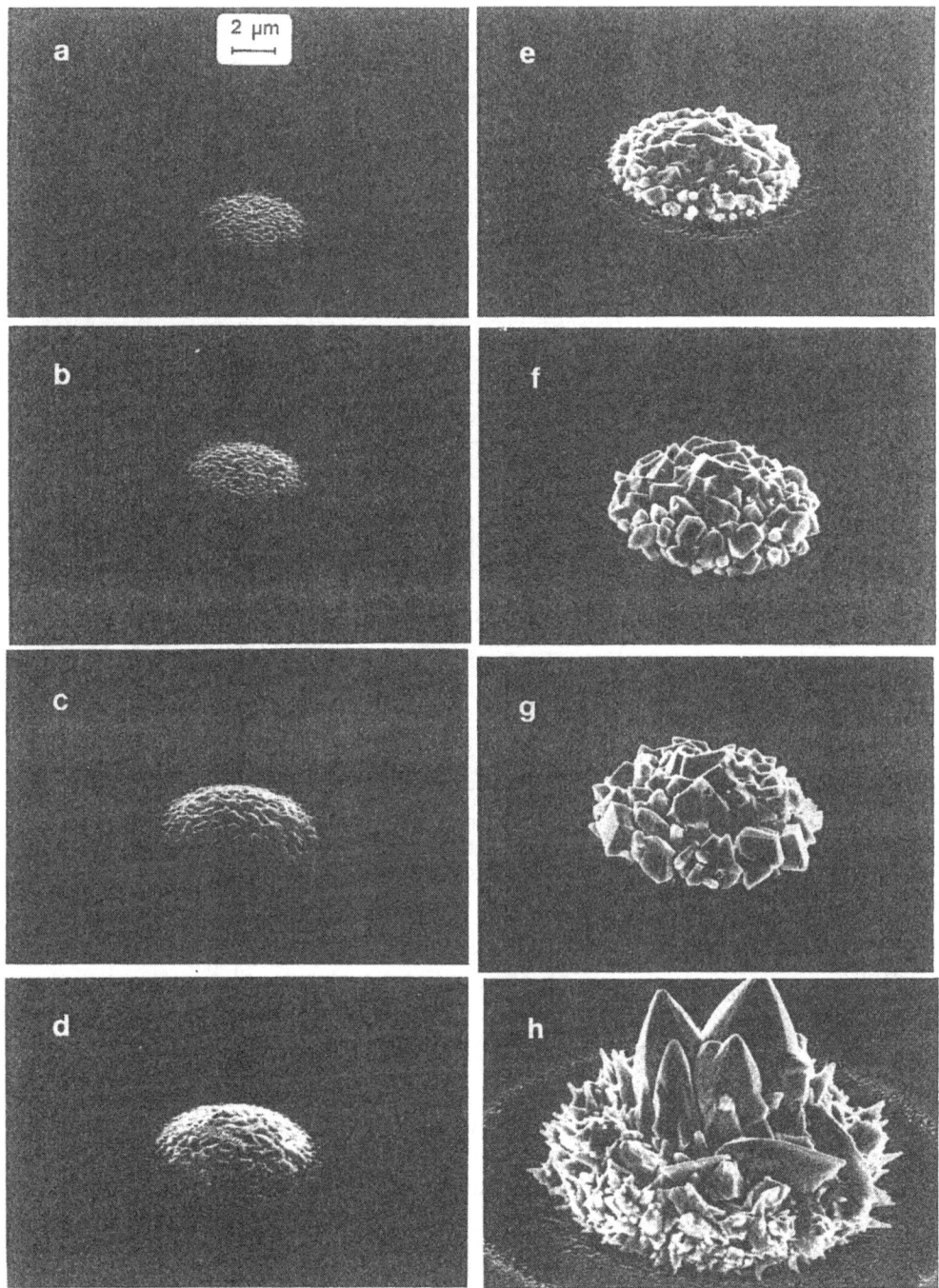

Fig. 3a–h. W spots deposited from an admixture of 5 mbar WF_6 and 500 mbar H_2 by means of 647.1 nm Kr^+ laser radiation ($P = 175$ mW, $2w_0 = 3.0$ μm). The substrate was fused quartz (a-SiO_2) covered with a 700 Å thick layer of sputtered W. The laser beam illumination time t_i was increased successively from a) to h) as follows: (a) 0.01 s. (b) 0.02 s. (c) 0.05 s. (d) 0.07 s. (e) 0.1 s. (f) 0.2 s. (g) 0.3 s. (h) 0.5 s (after [3])

of $WF_6 + H_2$ and both the laser beam and the substrate are at rest, one initially observes the deposition of a circular W spot whose diameter increases with increasing laser beam illumination time t_i. This is shown in Fig. 3 for a fused SiO_2 (a-SiO_2) substrate covered with a 700 Å thick layer of sputtered W. Figure 4 shows that the diameter of spots first increases very rapidly with t_i and then saturates at a certain value which depends on the laser power. Numerical differentiation of these results directly yields the lateral growth rate of spots $W_l(t_i) \equiv \Delta d(t_i)/2\Delta t$. From these growth rates and an analysis of the laser-induced temperature distribution,

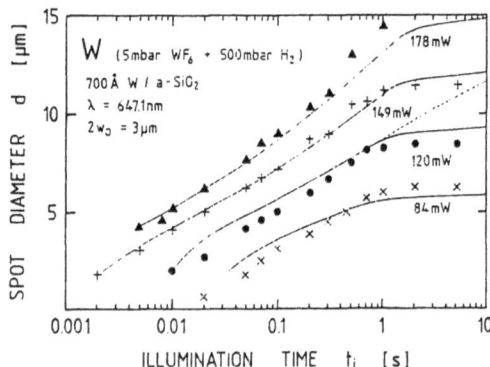

Fig. 4. Diameter of W spots for various laser powers as a function of 647.1 nm Kr$^+$ laser beam illumination time. The laser focus was $2w_0 = 3.0$ µm. The partial pressures of gases were 5 mbar WF$_6$ and 500 mbar H$_2$. The substrate was a-SiO$_2$ covered with 700 Å sputtered W. The full and dashed curves were calculated (after [3])

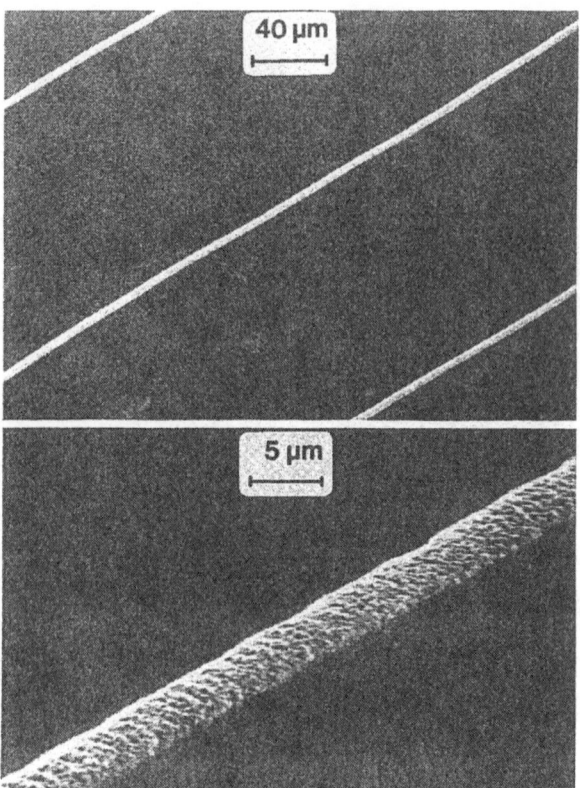

Fig. 6. Scanning electron micrographs of W stripes deposited on (100) Si wafers from a mixture of WF$_6$ and H$_2$ by means of 647.1 nm Kr$^+$ laser radiation (after [9])

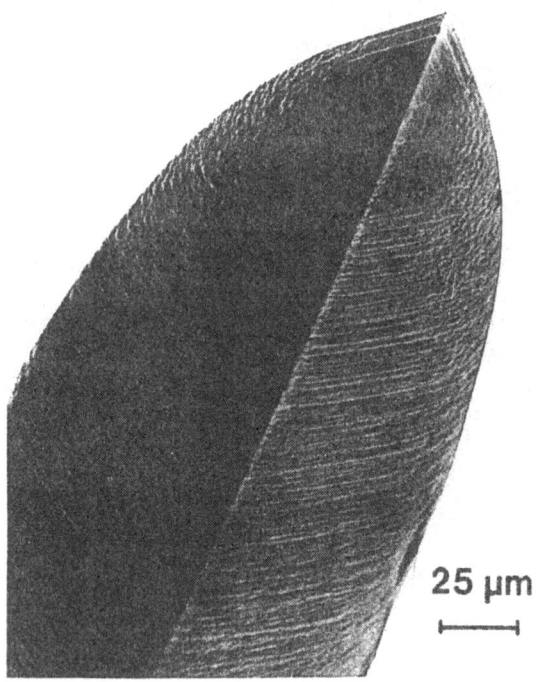

Fig. 5. Scanning electron micrograph of the tip of a single-crystal W rod grown from a mixture of WF$_6$ and H$_2$ by means of 488 nm Ar$^+$ laser radiation (after [7])

The saturation in the lateral growth rate of spots observed in Fig. 4 is related to the decrease in the laser-induced temperature rise at the edge of the spot. If irradiation is continued, the growth of a rod along the axis of the laser beam is observed. The microstructure of such rods can be amorphous, poly-crystalline or single-crystalline [4–6]. The tip of a laser-grown single crystal of W is shown in Fig. 5. Because the rods can be grown under quasi-stationary conditions, one can measure both the axial growth velocity and the temperature in the tip of the rod (via the emitted thermal radiation) with great accuracy. This is by far the best method for studying the reaction kinetics in pyrolytic LCVD [8]. It also allows for the in situ observation of localized high-temperature chemical reactions in otherwise cold environments. Therefore, such investigations yield valuable information on fundamental processes that are also relevant to technologies currently in widespread use, such as standard chemical vapour deposition (CVD).

In what follows we will briefly discuss pyrolytic laser direct writing of patterns. This is performed by moving the substrate perpendicularly to the laser beam (Fig. 2a). Some recent results are shown in Fig. 6 for the

one can directly derive the chemical kinetics that controls the deposition process, i.e. the apparent chemical activation energy, ΔE, and the preexponential factor in the Arrhenius type law for the reaction rate. With the present system this analysis yields $\Delta E = 30$ kcal/mol. Similar investigations have been performed with other substrate materials and precursor molecules.

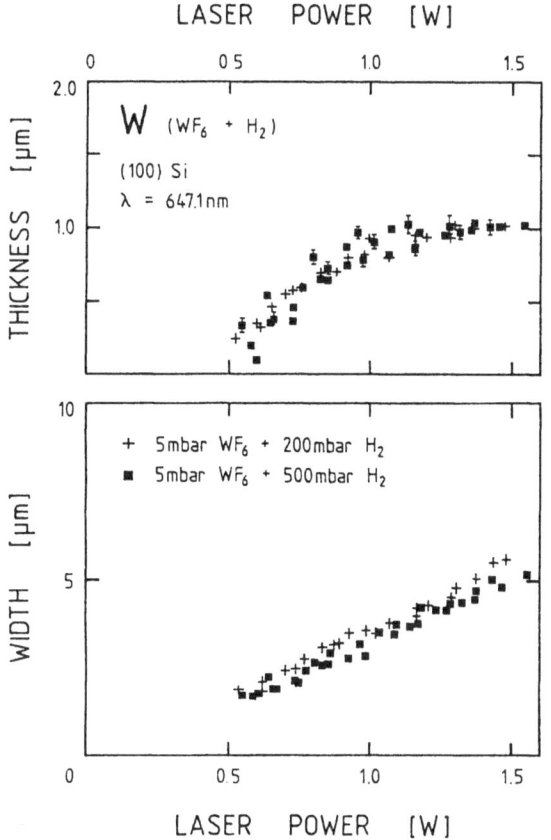

Fig. 7. Thickness and width of W stripes as a function of laser power. The substrate was (100) Si. The laser beam was focused a few microns *below* the substrate surface (after [9])

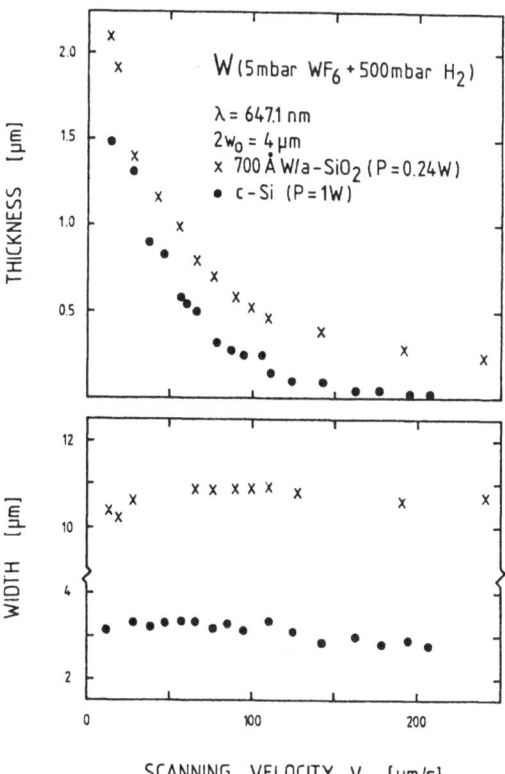

Fig. 8. Thickness and width of W stripes as a function of the scanning velocities of the laser beam. x focal plane on the substrate surface, ● focal plane within substrate surface (after [9])

example of W stripes deposited on (100) silicon wafers (c-Si). The microstructure of the stripes is polycrystalline. The grain size depends on the laser power, the partial pressures of gases, the scanning velocity etc. Figure 7 shows the thickness h and the width d of stripes as a function of laser power. In these experiments, the laser beam was focused a few microns *below* the Si surface. In this way, the formation of periodic structures [10] can be suppressed and the reproducibility of data thus significantly improved. Figure 8 shows the dependence of the thickness and width of stripes on scanning velocity for two different substrate materials. The decrease in thickness of stripes can be understood simply from the decrease in laser-beam dwell time $\tau_1 \sim 1/v_s$, while the explanation of the almost constant width of stripes is more complicated.

The resistance of stripes per unit length and the resistivity normalized to the bulk value of W ($\varrho_B = 5.33 \times 10^{-6}$ Ωcm) is shown in Fig. 9 as a function of laser power. The strong decrease in the resistance with increasing laser power is mainly due to the increase in the cross sections of stripes (Fig. 7), while the increase in the resistivity with increasing laser power is accounted for by stripe texture, which becomes more pro-

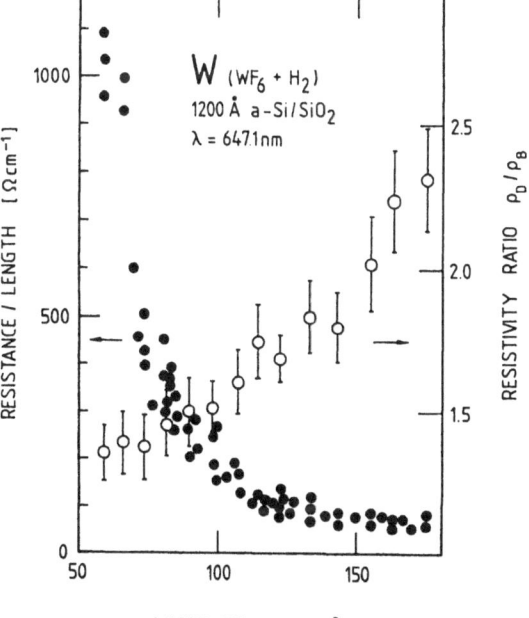

Fig. 9. Electrical resistance (left scale) and resistivity ratio (right scale) of W stripes as a function of laser power (647.1 nm Kr⁺, $2w_0 = 3.5$ μm). The partial pressures of WF$_6$ and H$_2$ were $p(\mathrm{WF}_6) = 5$ mbar and 100 mbar $\leq p(\mathrm{H}_2) \leq 800$ mbar. A value of $\varrho_B = 5.33 \times 10^{-6}$ Ωcm has been used for bulk W (after [11])

nounced the higher the laser power (the change in microstructure being similar to that shown in Fig. 3 for short and longer laser-beam illumination times).

Photolytic LCVD of microstructures in the form of spots and stripes was also investigated. In this case, deposition is based on the excitation of dissociative electronic transitions. Most of the experiments have been performed for the deposition of metals from the corresponding alkyls and carbonyls. These molecules show dissociative continua in the near to medium ultraviolet which can easily be reached with frequency-doubled ion lasers, with excimer lasers or with frequency-multiplied Nd:YAG lasers. While this technique is less sensitive to the physical properties of the substrate material, problems arise from the gas-phase nucleation and overall diffusion of photo-generated species. Additionally, the deposition rates achieved are several orders of magnitude lower than those achieved in pyrolytic LCVD.

Recently, various investigations have been performed to combine photolytic and pyrolytic processes. For example, KrF excimer laser projection has been used to photolyze adlayers of $Al(C_4H_9)_3$ on various substrate materials such as Al_2O_3, SiO_2, GaAs, and Si [12]. The reactive Al sites produced in this way served as nucleation centres for spatially selective growth of Al film patterns by standard CVD. An Al pattern produced in this way is shown in Fig. 10. The figure demonstrates that small bare areas surrounded by uniform deposits can be produced. In this technique, the resolution of features is controlled only by the imaging of the ultraviolet light and *not* by the gas-phase diffusion length. In other words, when e.g. ArF laser radiation is used together with proper imaging, it should be possible to improve the resolution by a factor of more than 20. The electrical resistivity of the

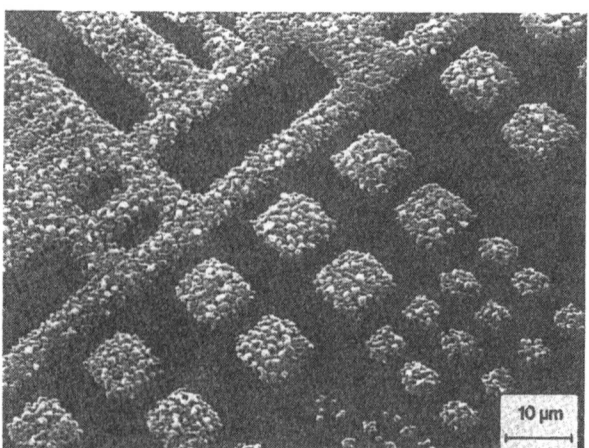

Fig. 10. Scanning electron micrograph of an Al pattern produced by photolysis of adsorbed $Al(C_4H_9)_3$ using KrF excimer-laser projection and subsequent standard CVD (after [12])

Al film shown in Fig. 10 is only about twice that of bulk Al. The technique has been used to produce metal patterns for metal oxide field-effect transistors and is compatible with standard photolithographic fabrication methods.

1.2. Thin Extended Films

For the deposition of thin extended films, the laser beam is used either at perpendicular and/or at parallel incidence to the substrate. For perpendicular incidence the situation is very similar to that discussed in the preceding section. The main difference is that gas- or liquid-phase transport limitations become important at lower (thickness) deposition rates. For *parallel* incidence the decomposition of molecules can again be based on either photothermal or photochemical excitation mechanisms which, in LCVD, are gas-phase heating and selective single or multiphoton excitation of electronic or vibrational transitions respectively. Clearly, species excited or dissociated within the volume of the laser beam are transported by diffusion and condense on the nearby substrate surface. Metal films have been deposited in this way mainly by gas-phase photolysis of metal-alkyls, carbonyls and fluorides. Various types of thin films of oxides and nitrides such as SiO_2, Al_2O_3, TiO_2, Si_3N_4, etc. have been deposited in a similar way by gas-phase photolysis of the appropriate precursor molecules. Deposition of thin extended films based on laser-induced gas-phase heating has been investigated in detail for the CO_2-laser-induced deposition of amorphous hydrogenated silicon (a-Si:H) films.

2. Surface Modifications of Materials

The laser-induced activation of a solid surface and/or an ambient medium can also result in a direct reaction between atoms or molecules (including photofragments) of the ambient medium, and atoms or molecules within the solid surface, or it may simply enhance the diffusion of species into or out of this surface. This is the case in laser-induced surface modification. Among the different types of surface modifications are: oxidation or nitridation of metals and semiconductors, laser-induced surface doping, and laser-induced surface reduction and metallization. The common feature in these different treatments is that the physical and/or chemical properties of the material surface are modified. Because surface modification, in general, requires thermal and/or non-thermal photoexcitations of the substrate, it is mostly performed at normal laser light incidence.

In the following we will only briefly mention some recent results on the laser-induced reduction of

$YBa_2Cu_3O_{7-x}$. In both the high-T_c superconducting phase ($0 \lesssim x \lesssim 0.2$) as well as in the semiconducting phase ($x \approx 1$), the oxygen content can be significantly diminished when a laser beam is scanned across or projected onto such material surfaces in a chemically reducing atmosphere such as hydrogen. The mechanism is based on the photothermally activated diffusion and removal of oxygen from the laser-processed surface regions. Because the electrical conductivity, including the superconductivity, depends so heavily on the oxygen content, the technique permits these physical properties to be locally changed. In fact, it has been demonstrated that the room temperature electrical conductivity can be locally enhanced by several orders of magnitude, depending on laser power, laser beam dwell time and H_2 pressure [13]. With a high degree of reduction, the laser-processed regions have a copper-like appearance. Laser-induced reduction and metallization have been studied in more detail for electro-optical PLZT ($Pb_{1-3y/2}La_yTi_{1-x}Zr_xO_3$) ceramics [14] and for single-crystal $BaTiO_3$ [15] and $LiNbO_3$ [16].

3. Compound Formation

Investigations have also been carried out into laser-induced alloying and compound formation from solid solutions or from alternating multiple layers consisting of appropriate proportions of the elements. Various types of silicides and even stoichiometric crystalline films of various compound semiconductors have been produced in this way [1].

4. Etching and Ablation

Many laser-induced chemical reactions and techniques discussed in the foregoing sections can be directly applied to the etching of materials. This has already been indicated for some model systems mentioned in Fig. 1, where in some cases the course of reaction can be reversed by simply shifting the chemical equilibrium to the other side. In fact, laser-induced chemical etching has also been demonstrated for a great variety of metals, semiconductors and insulators. In the following we will outline some recent results on the laser-induced chemical etching of silicon in a Cl_2 atmosphere and on the ablation of piezoelectric oxides and organic polymers. In contrast to laser-induced etching, laser-induced ablation can also be performed in a vacuum or a chemically inert atmosphere. With the materials discussed in this section, the microscopic mechanisms involved in the ablation process are, however, not simply based on laser-induced thermal evaporation alone.

4.1. Silicon

Laser-induced chemical etching of semiconductors is a powerful technique for processing electronic materials, the laser-induced dry-etching of silicon being of particular importance. Here, we will confine ourselves to outlining some recent results on the laser-induced etching of silicon in a chlorine atmosphere. Here, investigations employing pulsed laser radiation have shown that for laser fluences that cause negligible surface heating, laser-induced etching of Si requires both photo-generated carriers within the Si surface and Cl radicals produced by gas-phase dissociation of Cl_2 molecules. At medium laser fluences a photothermal activation of the etching process becomes important, but photo-generated Cl radicals are still required. At laser fluences that cause surface melting, photothermal activation of the etching process dominates [17].

In the following we will confine the discussion to the regime of low laser powers where etching is dominated by non-thermal (photochemical) microscopic mechanisms. The experiments employed both cw-laser irradiation and combined cw- and pulsed-laser irradiation. Further details will be published in [18, 19].

4.1.1. Continuous-Wave Laser Irradiation. The experiments involving only cw laser irradiation were performed by means of Ar^+ and Kr^+ laser radiation at wavelengths between about 351 nm and 647 nm. The laser beam was focused in each case to a spot of $2w_0 = 30 \mu m$ diameter ($1/e^2$ intensity). The samples were slightly p-doped (100) Si wafers ($\varrho = 100$–$150 \Omega cm$). Typically, holes of 10 μm to 15 μm depth were etched.

Figure 11 shows the etch rate as a function of Cl_2 pressure at various laser wavelengths. At fixed laser power and laser wavelength, the generation rate of photo-carriers within the silicon surface is constant and the change in etch rate with Cl_2 pressure is controlled by the change in the Cl atom concentration above the Si surface. Various techniques can be employed to measure the Cl atom concentration but they require a considerable expenditure of time and effort and/or are unable to deliver a spatial resolution of better than about 30 μm which is the minimum required in the analysis of the present experiments. Therefore, we have calculated the flux of Cl atoms to the surface. The solid and dashed curves in Fig. 11 show the results of the numerical calculations. The etch probability η_0 has been used as a fit parameter by taking into account the proportionality between the net flux of chlorine atoms j_{Cl} and the etch rate W. The agreement between the experimental data and the calculated curves is excellent. Both the kinetically

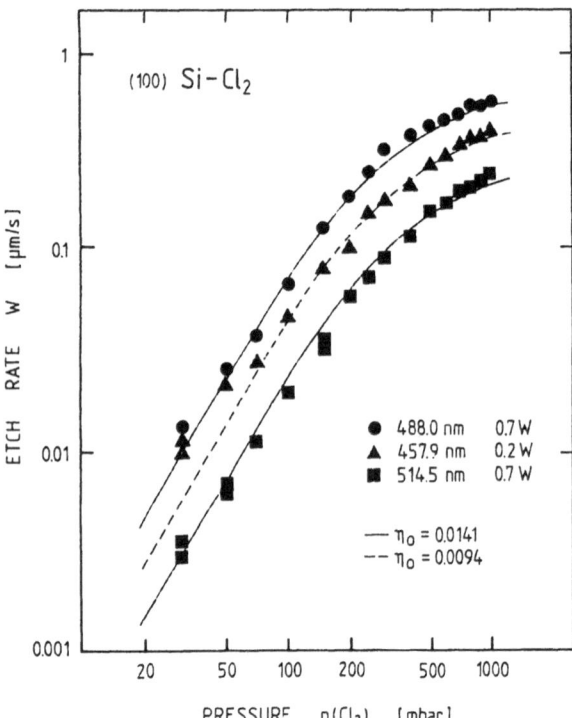

Fig. 11. Etch rate for holes in (100) Si surfaces as a function of Cl_2 pressure for three different Ar^+ laser wavelengths. The focus was $2w_0 = 30\,\mu m$ in all cases. Note that the curve with $\lambda = 457.9\,nm$ was obtained with a laser power of $P = 0.2\,W$, while the curves at $\lambda = 488.0\,nm$ and $514.5\,nm$ were taken with $P = 0.7\,W$. The solid and dashed curves have been calculated (after [18])

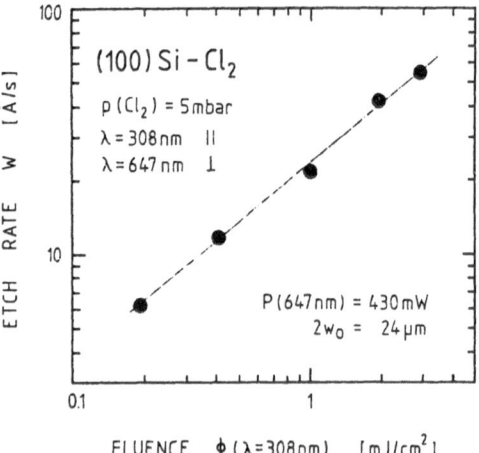

Fig. 12. Etch rate W as a function of the excimer laser fluence Φ at parallel incidence to the substrate, at constant Kr^+ laser power perpendicularly incident to the substrate surface. The solid line is a fitted curve showing a slope of 0.80 (after [19])

controlled regimes and the mass transport limited regimes are described equally well, showing that it is indeed the chlorine atoms that are involved in the etching process.

The curves obtained for 488.0 nm and 514.5 nm radiation can be fitted with the same value of η_0. This is not surprising, as these curves were obtained at the same laser power, and a calculation of the laser-induced photoelectron density gives almost identical dependencies for these two wavelengths. On the other hand, at 457.9 nm the curve can be fitted only by using a lower value of η_0 which corresponds to the lower laser power used in the experiments. Here one should note that a change in laser power P results in a change in both the Cl atom concentration in the gas phase and the density of photoelectrons within the silicon surface and, consequently, in a change of the reaction probability η_0.

4.1.2. Combined cw and Pulsed Laser Irradiation. In this irradiation scheme the generation of Cl atoms in the gas phase and the generation of photoelectrons in the Si surface is performed separately by means of two different laser beams. The Cl atoms are generated by a XeCl excimer-laser beam at parallel incidence to the

sample surface at a distance of several millimetres. The resulting Cl atom concentration as a function of Cl_2 pressure and laser fluence was monitored by detecting the chemiluminescent radiation at a wavelength of $\lambda = 550\,nm$ resulting from the Cl–Cl atom recombination [19]. The silicon sample was irradiated simultaneously with a 647 nm Kr^+ laser beam which was focused to a spot of $2w_0 = 24\,\mu m$ diameter. At 647 nm the absorption of Cl_2 gas is negligible. This means that the chlorine atoms are only generated by the ultraviolet excimer-laser radiation, while the photo-carriers are only generated by the red Kr^+ laser radiation. The etched holes were rather flat with typical depths of about 1 μm only. The diameter of holes was in all cases larger than the diameter of the laser spot and it is in good agreement with the calculated spatial distribution of photoelectrons.

Figure 12 shows the dependence of the etch rate on the excimer-laser fluence Φ. The observed dependence $W \sim \Phi^{0.80}$ is in good agreement with the chlorine atom density derived from chemiluminescence experiments $N_{Cl} \sim \Phi^{0.78}$, and the calculated dependence $N_{Cl} \sim \Phi^{0.73}$. The proportionality observed between the etch rate and the chlorine atom concentration is further proof of the fact that it is Cl atoms which are the reactive species in the photolytic etching of silicon.

4.2. Oxides

Various types of perovskites and perovskite-related oxides have been etched by means of cw laser irradiation in a H_2 atmosphere [1, 20]. In the following we will report on the ablation of these materials by excimer laser projection. Figure 13 shows a scanning

Fig. 13. Hole produced in LiNbO$_3$ by XeCl excimer-laser irradiation with 500 pulses of $\Phi = 2.7$ J/cm^2 in a vacuum of 10^{-5} mbar. The diameter of the beam spot on the sample surface was $2w = 175$ μm (full width half maximum) (after [21])

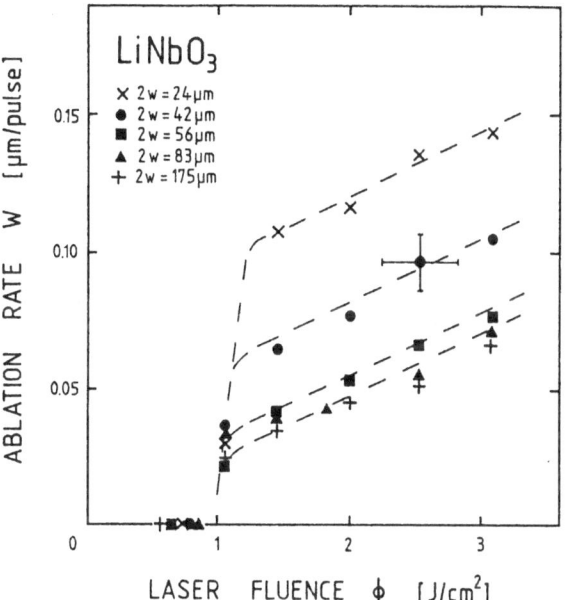

Fig. 14. Ablation rate W for XeCl excimer-laser-irradiated LiNbO$_3$ as a function of laser fluence for various spot diameters. The dashed lines are guides for the eye (after [21])

electron micrograph of a hole produced in single-crystal LiNbO$_3$ [21]. The ablation process can be seen to produce a very regular and smooth surface at the bottom of the hole. No evidence of damage can be seen in the material adjacent to the hole, apart from an approximately 1 μm thick brittle layer surrounding the hole. Structures of similar quality have also been produced in single-crystal BaTiO$_3$. Figure 14 shows ablation rates for LiNbO$_3$ as a function of laser fluence and for various laser-beam spot sizes produced by inserting apertures of various diameters into the beam path (Fig. 2b). Note that the ablation rates are con-

siderably higher than those obtained in ion-beam milling, which is the conventional method for structuring LiNbO$_3$. We have not observed any effect on the ablation rates when working either in air or vacuum, although the surface quality of the structures achieved is somewhat better in vacuum. Above a threshold fluence Φ_d, surface damage is induced in these materials without significant ablation. Then, above a second threshold of $\Phi_{th} \approx 0.9$ J/cm^2 rapid ablation occurs in both LiNbO$_3$ and BaTiO$_3$ with ablation rates of nearly equal magnitude. The similarity in ablation rates and threshold fluences observed for LiNbO$_3$ and BaTiO$_3$ is quite surprising and cannot be explained on the basis of a purely thermal mechanism. This is because the absorption coefficients for LiNbO$_3$ and BaTiO$_3$ differ by several orders of magnitude. Even when the strong temperature dependence of the absorption coefficient of LiNbO$_3$ is taken into account, the laser-induced temperature rise calculated for a fluence of $\Phi = 1$ J/cm^2 and a pulse length of 11 ns should be more than 30 times higher for BaTiO$_3$ than for LiNbO$_3$. This is in contrast to the measured ablation rates which are about the same for both materials. Therefore, nonthermal mechanisms which may include dense carrier excitations, as previously discussed for semiconductors and oxides [22, 23], as well as laser-light – surface-plasma interactions, seem to be involved in the etching mechanism.

Another important feature observed in Fig. 14 is the pronounced dependence of the ablation rate on the laser-beam spot size. The ablation rates are higher for smaller spot sizes than for larger ones, once the spot size is decreased below the saturation value of about 80 μm. We have evidence that a dependence of the ablation rate on the beam spot size is quite a general phenomenon in pulsed laser-induced material ablation. We believe that this effect originates from the shielding of the incident laser radiation. This shielding decreases with decreasing laser-beam spot sizes because then 3-dimensional particle emission becomes effective. This is consistent with time-dependent reflectivity measurements. From a practical point of view, this effect has to be considered when one is operating with projection masks with features of different dimensions. In this case one has to expect different ablation rates!

4.3. Organic Materials

An example of laser light breaking chemical bonds *directly* within the surface of the material, is the ablation of organic polymers and biological tissues by means of ultraviolet excimer-laser radiation [24]. Figure 15 shows etch patterns obtained in a multilayer photoresist [25]. The resolution achieved is 0.35 μm.

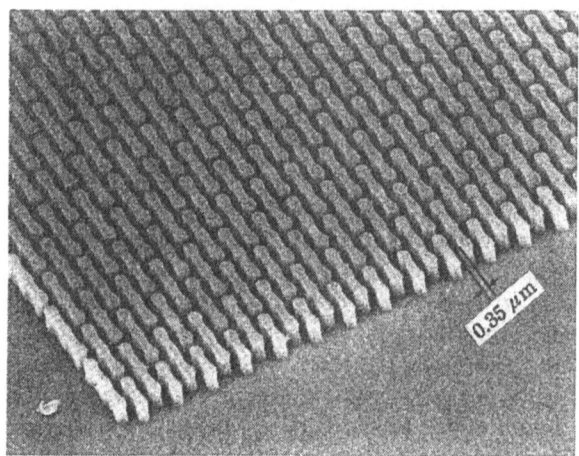

Fig. 15. Pattern produced in photo-resist by 10:1 refractive optics, using a KrF laser as light source. The resist consists of multiple-layers of PMMA (1 μm)/SOG (0.5 μm)/OFPR 800 (1.2 μm) (after [25])

Because ablation can be performed in a vacuum or in a non-absorbing atmosphere, high laser irradiances can be used. As a consequence, much higher rates of material removal can be obtained than those achieved in photolytic etching from gas- or liquid-phase precursors. The ablation rates range, typically, from some 0.1 μm/pulse up to several μm/pulse.

5. Summary

Laser-induced chemical reactions at or near solid surfaces offer manifold possibilities for processing materials. In laser pyrolysis the reaction rates are, in most cases, several orders of magnitude higher than those in photolysis. The higher reaction rates allow for higher scanning speeds in direct writing and, in the case of deposition, the production of three-dimensional structures. In addition, the microstructures and the electrical properties of pyrolytically deposited films are in general superior. In many cases etching reactions can be localized better in pyrolysis than in photolysis.

Laser-induced etching of slightly p-doped silicon in a chlorine atmosphere requires both Cl atoms in the gas phase and photo-carriers within the Si surface. The etch rate is proportional to the Cl atom concentration above the surface.

Laser-induced surface reduction of perovskites and perovskite-related oxides permits single-step direct writing of conducting patterns within the material surface. The (normal) electrical conductivity of the substrate surface can be locally enhanced by many orders of magnitude. Patterning of these materials by excimer-laser-induced ablation yields smooth surfaces and negligible damage to the surrounding material. The ablation rates achieved depend on the laser-beam spot size.

Acknowledgements. I wish to thank Drs. M. Eyett, G. Higashi, Y. Horiike, R. Kullmer, K. Piglmayer, T. Szörényi, and G. Q. Zhang for permission to include their results in this review. I also wish to thank the "Fonds zur Förderung der wissenschaftlichen Forschung in Österreich" for financial support.

References

1. D. Bäuerle: *Chemical Processing with Lasers*, Springer Ser. Mater. Sci. **1** (Springer, Berlin, Heidelberg 1986)
2. D. Bäuerle (ed.): *Laser Processing and Diagnostics*, Springer Ser. Chem. Phys., Vol. **39** (Springer, Berlin, Heidelberg 1984)
3. T. Szörényi, K. Piglmayer, G.Q. Zhang, D. Bäuerle: To be published
4. D. Bäuerle: In *Laser Diagnostics and Photochemical Processing for Semiconductor Devices*, ed. by R.M. Osgood, S.R.J. Brueck, H.R. Schlossberg (North-Holland, New York 1983) p. 19
5. D. Bäuerle: Laser Optoelektron. **1**, 29 (1985)
6. D. Bäuerle: In *Laser Processing and Diagnostics*, Springer Ser. Chem. Phys., Vol. **39** (Springer, Berlin, Heidelberg 1984) p. 166
7. T. Szörényi, M. Boman, D. Bäuerle: Unpublished
8. J. Doppelbauer, D. Bäuerle: In *Interfaces under Laser Irradiation*, ed. by L.D. Laude, D. Bäuerle, M. Wautelet, Nato ASI Series (Nijhoff, Dordrecht 1987) p. 277
9. T. Szörényi, G.Q. Zhang, D. Bäuerle: Unpublished
10. T. Szörényi, G.Q. Zhang, Y.C. Du, R. Kullmer, D. Bäuerle: In *Laser Processing and Diagnostics II*, ed. by D. Bäuerle, K.L. Kompa, L.D. Laude (Physique, Les Ulis 1986) p. 91
11. G.Q. Zhang, T. Szörényi, D. Bäuerle: J. Appl. Phys. **62**, 673 (1987)
12. G.S. Higashi, G.E. Blonder, C.G. Fleming: In *Photon, Beam, and Plasma Stimulated Chemical Processes at Surfaces*, ed. by V.M. Donnelly, I.P. Herman, M. Hirose (Materials Research Society 1987) p. 117
13. G. Liberts, M. Eyett, D. Bäuerle: Appl. Phys. A **45**, 313–316 (1988)
14. A. Kapenieks, M. Eyett, D. Bäuerle: Appl. Phys. A **41**, 331 (1986)
15. U. Kolzer, M. Eyett, D. Bäuerle: To be published
16. M. Eyett, R. Kullmer, D. Bäuerle: In *Proc. SPIE Int. Symp. on Optical and Optoelectronic Applied Science and Engineering, Topical Meeting on High Power Lasers*, Vol. **801** (1987) p. 156
    M. Eyett, D. Bäuerle: Ferroelectrics Lett. (1988) (to be published)
17. R. Kullmer, D. Bäuerle: Appl. Phys. A **43**, 227 (1987)
18. P. Mogyorósi, K. Piglmayer, R. Kullmer, D. Bäuerle: Appl. Phys. A **45**, 293–299 (1988)
19. R. Kullmer, D. Bäuerle: Appl. Phys. A **43**, 227–232 (1987)
20. M. Eyett, D. Bäuerle, W. Wersing, H. Thomann: J. Appl. Phys. **62**, 1511 (1987)
21. M. Eyett, D. Bäuerle: Appl. Phys. Lett. **51**, 2054 (1987)
22. M. Wautelet, L.D. Laude: Appl. Phys. Lett. **36**, 197 (1980)
23. N. Itoh, T. Nakayama: Phys. Lett. **92**A, 471 (1982)
24. R. Srinivasan: In *Laser Processing and Diagnostics*, ed. by D. Bäuerle, Springer Ser. Chem. Phys., Vol. **39** (Springer, Berlin, Heidelberg 1984) p. 343; In *Interfaces under Laser Irradiation*, ed. by L.D. Laude, D. Bäuerle, M. Wautelet, Nato ASI Series (Nijhoff, Dordrecht 1987) p. 359
25. Y. Horiike, N. Hayasaka, M. Sekine, T. Arikado, M. Nakase, H. Okano: Appl. Phys. A **44**, 313 (1987)

Appl. Phys. B 46, 271–282 (1988)

Applied Physics B

Photo-
physics
and Laser
Chemistry

© Springer-Verlag 1988

Laser Spectroscopy Applied to Energy, Environmental and Medical Research

S. Svanberg

Department of Physics, Lund Institute of Technology, P.O. Box 118,
S-22100 Lund, Sweden

Received 10 March 1988/Accepted 14 March 1988

Abstract. Through the development of powerful laser spectroscopy techniques new means for advanced diagnostics and sample analysis have emerged. Applications of laser spectroscopy in the fields of energy, environmental and medical research are discussed. Emphasis is placed on non-intrusive diagnostic techniques for studying combustion processes, for remote monitoring of atmospheric pollutants and for industrial and medical applications of laser-induced fluorescence. Selected examples from work performed at the Lund Institute of Technology are used as illustrations, and references to books, reviews and selected papers are given.

PACS: 82.80, 87.00, 82.50

The development of tunable high-intensity light sources, in particular the tunable dye laser, in conjunction with the introduction of powerful laser spectroscopic techniques has had a strong influence on optical spectroscopy. In the field of *basic research* a vast amount of new data on the static and dynamic properties of atomic and molecular systems have been obtained. New *fundamental* experiments for precision determinations of natural constants, for tests of parity violation and the isotropy of space, have been designed. Recent progress in these areas is covered in the proceedings of the latest international conferences of laser spectroscopy [1, 2] and atomic physics [3, 4].

Applied laser spectroscopy frequently means new analytical and diagnostic techniques, which have opened up new vistas in many different disciplines such as physics, chemistry, biology and medicine. In particular, many cross-disciplinary activities on the borderline between traditional fields have emerged. Common and attractive features of laser-spectroscopic diagnostic techniques are:

* High sensitivity
* High selectivity
* Non-intrusiveness
* Near real-time analysis.

Analytical laser spectroscopy techniques are rapidly expanding the frontiers of chemical analysis [5].

New insight into the dynamics of chemical reactions is being gained and the course of the reactions can be influenced. Laser-induced chemistry provides new production techniques for special attractive chemicals, and isotopic selectivity can even be gained, providing new schemes for isotope separation [6, 7]. Laser diagnostics and photochemical processing are also gaining in importance in connection with semiconductor devices [8–10]. Many aspects of basic and applied activities in laser spectroscopy are discussed in [11, 12].

The present paper on applied laser spectroscopy will focus on some applications in energy, environmental and medical research. Three specific fields are discussed: combustion diagnostics, remote sensing of atmospheric pollution, and diagnostics of cancer tumours and atherosclerotic plaque. Illustrations from work performed at the Lund Institute of Technology are chosen and general references to books and review articles covering the field are given.

1. Combustion Diagnostics

One often associates laser-driven fusion with laser applications in energy research. However, another important energy-related application of lasers is combustion diagnostics with the goal of a better under-

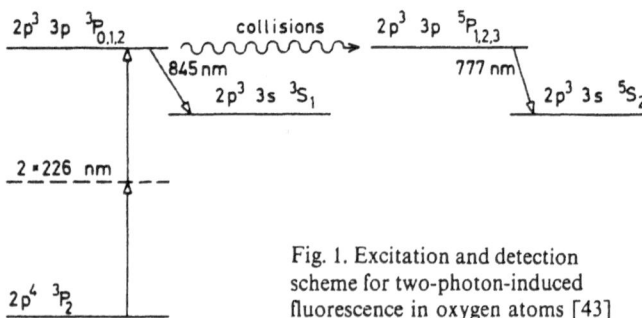

Fig. 1. Excitation and detection scheme for two-photon-induced fluorescence in oxygen atoms [43]

standing of combustion processes allowing improvement of fossil fuel combustion efficiency and reduction of pollution. Combustion technology is a very mature although empirical activity. In view of the increasing demands of high-efficiency energy conversion in an environmentally acceptable way it becomes very clear, that a detailed understanding of combustion on the molecular level is required [13, 14]. During recent years many useful diagnostic techniques for combustion studies have been developed based on laser spectroscopy. Concentrations of molecules, radicals and atoms can be measured and temperatures be determined by analysing signal intensities and distributions. In addition, gas flows and turbulence can be studied using Doppler shifts or spectral tagging of gas packets. General descriptions of the field of laser combustion diagnostics are found in [15–19]. Among the laser techniques useful for combustion diagnostics the following can be mentioned:

* Absorption spectroscopy (including tomography) [20, 21]
* Laser-induced fluorescence (LIF) [22–26]
* Rayleigh and Mie Scattering [27]
* Raman Scattering [28, 29]
* Coherent Anti-Stokes Raman Scattering (CARS) [30–32] and other coherent Raman techniques
* Degenerate four-wave mixing [33, 34]
* Photodeflection spectroscopy (including tomography) [35–37]
* Photoacoustic spectroscopy [38, 39]
* Optogalvanic spectroscopy (40–42].

Some relevant references for each of the techniques are indicated.

Among these techniques some comments will be made on laser-induced fluorescence (LIF), which is a powerful and very versatile technique for studies of species concentrations, flow and temperature fields. Single-step, two-photon or multiphoton transitions can be utilized to induce the fluorescence. Measurements can be performed in basically two different ways. The induced fluorescence light from a chosen spatial location can be spectrally analysed with a scanning monochromator or with an optical multichannel analyser system. Alternatively, the streak of LIF through

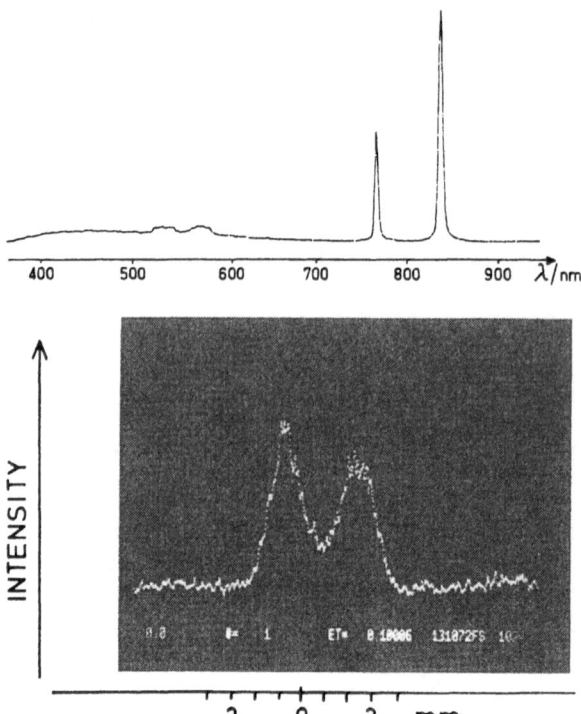

DISTANCE ACROSS THE FLAME

Fig. 2. (a) Fluorescence light recording for oxygen atoms in an acetylene-oxygen flame. (b) Single-shot imaging recording of the distribution of oxygen atoms in an acetylene-oxygen flame [43]

the flame can be imaged by a lens through an appropriate filter onto the diode array of the optical multichannel system. In both cases, gated electronics ensures the rejection of background light. In this way, single-shot spectral recording of turbulent flames or explosions become possible.

As illustrations of these two modes of measurement we chose studies of free oxygen atoms in an acetylene-oxygen welding torch [43]. An excitation and detection scheme for two-photon-induced fluorescence is shown in Fig. 1. Single-photon excitation is not possible in this case since the resonance transition of oxygen falls in the vacuum-UV part of the spectrum. By using two 226-nm photons the large energy gap of 10.6 eV up to the $2p^3 3p\ ^3P$ state can be bridged. The laser radiation is generated by frequency doubling of the output of a tunable dye laser and then mixing this light with residual 1.064 μm radiation from the pumping Nd:YAG laser. Fluorescence light released in the decay of the excited oxygen atoms to the $2p^3 3s\ ^3S_1$ state is observed at 845 nm as shown in Fig. 2a. In this recording, obtained with an optical multichannel analyser, strong fluorescence at 777 nm can also be observed. Near-resonant collisional transfer to the quintet manifold is responsible for this emission. In Fig. 2b a single-shot imaging recording of the streak of laser-induced fluorescence through the welding torch flame is shown, revealing the spatial location of the oxygen atoms during this 5 ns exposure.

Temperature can be measured both point-wise and in an imaging mode by observing fluorescence induced from thermally differently populated levels in naturally occurring molecules such as OH or seeded atoms such as In [44–46].

Flow velocity measurements using LIF are based on the shift of the Doppler-broadened absorption profile of atoms or molecules following the flow. The principle is illustrated in Fig. 3 [47]. To the right in this figure a recording of the spectral profile of an individual ro-vibronic line of I_2 molecules following a flow is given, recorded employing a cw single-mode dye laser. For flow measurements the laser is tuned halfway up the static-gas line profile, as indicated in the recorded profile, and further detailed in the left part of the figure. A certain fluorescence intensity is recorded. If the gaseous medium starts flowing the whole line profile is shifted to higher or lower frequencies depending on the direction of the flow and the fluorescence intensity is increased or decreased correspondingly. By taking the difference of the recorded intensities for laser beams propagating in opposite directions through the medium and normalizing to the sum of these intensities, as indicated in the formula included in Fig. 3, the velocity can be inferred without any knowledge of the concentration of the molecules under study. Imaging recordings of the intensity of iodine fluorescence for both laser-beam directions are shown in Fig. 4 for the case of an air jet investigated at an angle of 10° from the jet flow direction. The influence of the jet on the fluorescence recordings is clearly seen. In the figure the image of the velocity profile through the jet is also included as calculated form the fluorescence curves for each spatial location. These data were recorded at low pressures where collisional quenching is not severe. At higher pressures the signal is reduced and pulsed laser excitation must be employed in order to saturate the transition and keep the fluorescence intensity up.

Imaging two-dimensional measurements of species concentrations, temperature and flow fields are now being performed by several groups. Particularly extensive work is being pursued at Stanford University [24]. Single-photon, two-photon or stepwise excitation can be used to investigate a large variety of species.

Coherent Anti-Stokes Raman Scattering (CARS) provides strong signals and excellent background light rejection which makes it particularly attractive for studying real-world systems such as car engines [48] or industrial burners [49]. CARS is much used for temperature determinations, generally utilizing the nitrogen molecule, which is abundant in most combus-

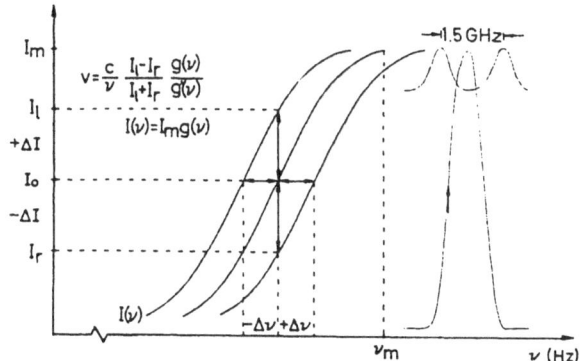

Fig. 3. Principle of velocity measurement using a single-mode laser tuned to the wing of a molecular absorption line [47]

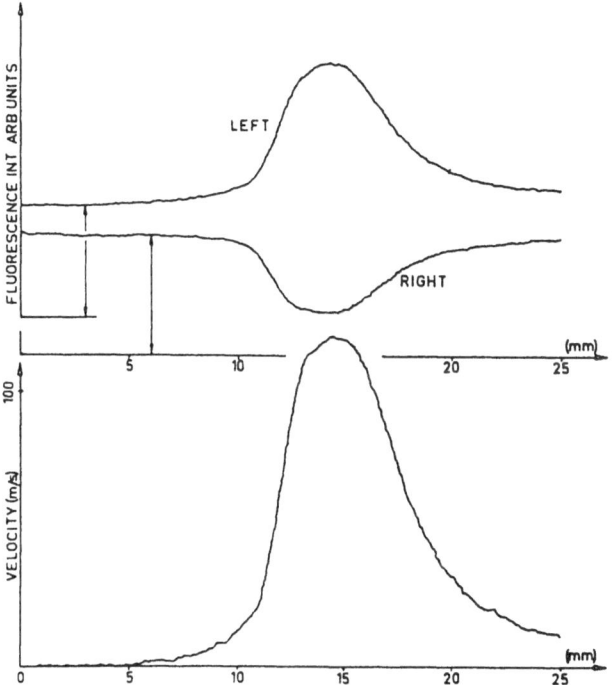

Fig. 4. Imaging recordings of the fluorescence light intensity along a line crossing an iodine-seeded jet for two directions of the single-mode laser beam. At the bottom of the figure the velocity profile of the jet is evaluated using the formula in Fig. 3 [47]

tion systems. The sensitivity of the CARS technique for species concentrations is much lower than that of LIF. The broadband CARS technique, providing the extraction of the relevant data from a single laser shot is of particular interest. Recent developments include the introduction of new multiplex techniques for multi-species monitoring and the increased utilization of pure rotational CARS [50]. Further efforts aim at the full understanding of CARS noise generation in view of the mode structure of the lasers used to ensure a high accuracy in the measurements [51].

274

2. Environmental Remote Sensing

Advanced techniques are needed to monitor our threatened environment, to evaluate pollution levels and developmental trends. While tropospheric pollution has obvious manifestations in terms of health problems, water and soil acidification, and forest damage, human-induced stratospheric changes in the ozone layer, as evidenced by the occurrence of "ozone holes" at the polar caps, may have much more far-reaching consequences [52–55]. Laser spectroscopy provides powerful means for remote sensing of molecules in the atmosphere, yielding information on pollution levels as well as meteorological conditions. There are two major kinds of laser methods applicable in remote sensing [56–60]:

* Long-path absorption monitoring
* Lidar (Light detection and ranging) with subdivisions:
  Fluorescence lidar
  Raman scattering lidar
  Mie scattering lidar
  Differential absorption lidar (dial).

Long-path absorption techniques are based on the same principles as spectrophotometry. However, by using laser beams of low divergence it is possible to use a pathlength of several km instead of the 1 cm cuvette used in the chemical laboratory. A single-ended arrangement can be achieved by utilizing a corner cube retroreflector at the end of the light path and collecting the back-reflected light with a telescope. Tunable diode lasers and cw line-tunable CO_2 lasers are useful in this approach. Since all detected photons have travelled the same path, no range resolution is obtained and only average concentrations can be determined. The technique has a powerful non-laser counterpart in doas (differential optical absorption spectroscopy) [61, 62], where a distinct high-pressure xenon lamp is used in combination with fast spectral scanning detection to overcome limitations posed by atmospheric scintillation.

In the lidar approach a laser pulse is transmitted into the atmosphere and backscattered radiation is detected as a function of time by an optical receiver in a radar-like fashion. In the case a fluorescence lidar a laser tuned to the absorption line of an atmospheric species is used and fluorescence light is detected in the subsequent decay. Fluorescence lidar is a powerful technique for measurements at mesospheric heights where the pressure is low and the fluorescence is not quenched by collisions. The technique has been used extensively to monitor layers of various alkali and alkaline earth atoms (Li, Na, K, Ca, and Ca^+) at a height of about 100 km [63]. Raman lidar has many attractive features but suffers from severe restrictions due to the weakness of the Raman scattering process. Only high gas concentrations can be detected at short ranges using high-power lasers and optical detection in a passband corresponding to the particular Raman shift. Water vapour profiles can be obtained in vertical soundings up to several km in height and pressure profiles up to tens of km are measurable using Raman signals from atmospheric N_2.

Mie scattering from particles provides strong signals allowing mapping of the relative distribution of particles over large areas. Stratospheric dust from volcanic eruptions can also be studied [64]. However, since Mie scattering theory involves many normally inaccessible particle parameters, quantitative results are difficult to obtain. Mie scattering is, however, extremely useful in providing the "distributed mirror" needed in differential absorption lidar (dial). The principles for dial are schematically represented in Fig. 5 [65]. Laser light is alternately transmitted at a wavelength where the species under investigation absorbs, and at a neighbouring, off-resonant wavelength. In the presence of an absorbing gas cloud the on-resonance signal is attenuated through the cloud and the off-resonant one is not. By dividing the two lidar signals by each other, most troublesome and unknown parameters are eliminated and the gas concentration as a function of the range along the beam can be evaluated. The dial technique is operational for important pollutants such as SO_2, NO_2,

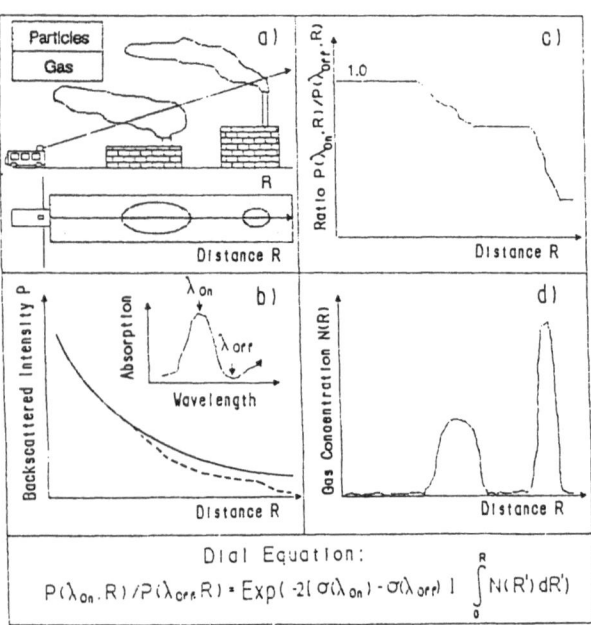

Fig. 5a–d. Illustration of the principle of differential absorption lidar (dial). (a) Pollution measurement situation, (b) backscattered laser intensity for the on- and off-resonance wavelengths, (c) ratio (dial) curve, (d) evaluated gas concentration [65]

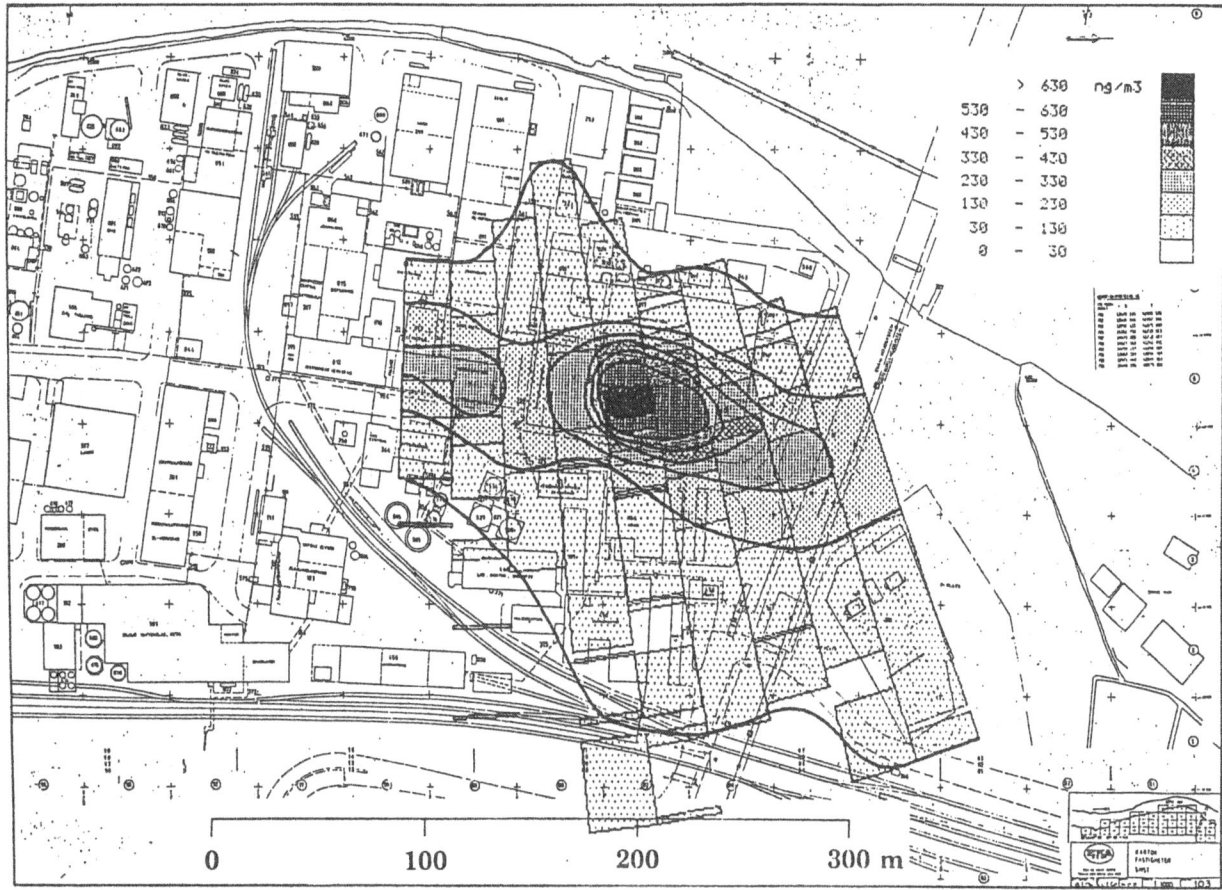

Fig. 6. Horizontal scan of Hg atom distribution over a chlorine-alkali plant [67]

and O_3 (see, e.g. [66]) and demonstrations for NO, HCl, Hg etc. have been performed.

As an example, the results of a dial measurement on a spreading plume of Hg atoms from a chlorine-alkali plant are shown in Fig. 6 as calculated directly by the system computer of a lidar system sweeping its laser beam horizontally over the plant [67]. Mercury is a troublesome pollutant but also an interesting geophysical tracer gas related to the occurrence of ore deposits, geothermal reservoirs and seismic activity [68–70].

Using ground-based dial systems the stratospheric ozone layer is being studied in the search for anomalies and long-term trends [71–73]. Both NASA and ESA are planning space lidar systems providing global wind, temperature and pollution monitoring.

Lidar systems are very powerful and provide unique three-dimensional information on atmospheric conditions. However, the systems also tend to be complex and expensive, which is a limiting factor in the widespread use of such systems for environmental management. Thus, there is a great need to develop simplified equipment. One such approach, gas correlation lidar, is illustrated in Fig. 7 [74]. Here a rather

crude laser system with a comparatively broad line-width is utilized. Since the laser wavelength is not sharp it covers both on- and off-resonance wavelengths at the same time, as illustrated in the figure. However, the information can be separated on the detection side by splitting the received radiation into two parts. One part is detected directly while the other part is first passed through a cell filled with an optically thick sample of the gas to be studied. In this way all the on-resonance radiation is filtered away leaving only the off-resonance radiation to be detected. In the direct channel the sum of the on- and off-resonance radiation is detected. By dividing the signals unknown factors are eliminated. The simultaneous detection of the two signals also eliminates influences due to atmospheric turbulence and fluctuations due to changing reflectivity in airborne measurements using topographic targets.

Hydrospheric pollution monitoring can be performed with airborne laser-based fluorosensors. A schematic diagram of a system for remote monitoring of laser-induced fluorescence is shown in Fig. 8 [75]. Different kinds of oil can be identified by their

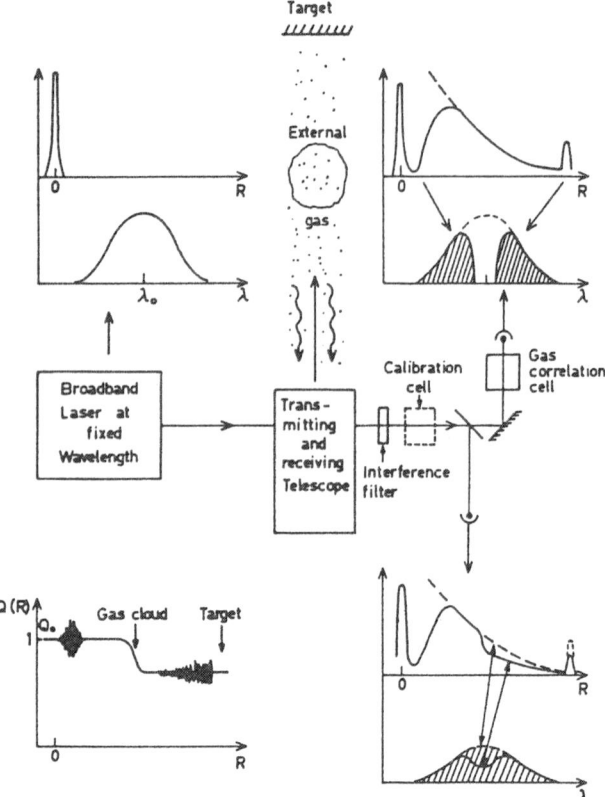

Fig. 7. Principle of the gas correlation lidar approach [74]

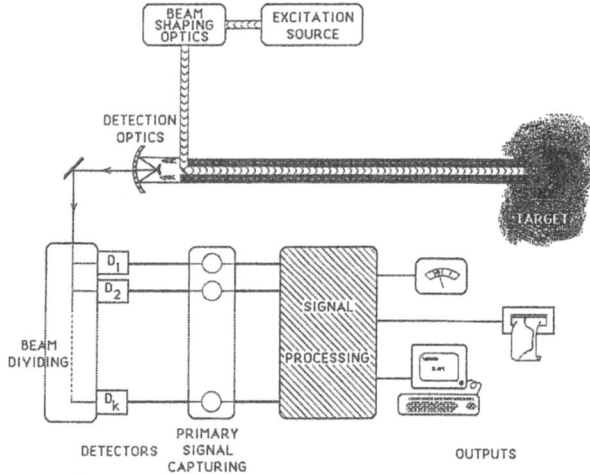

Fig. 8. Set-up for remote sample characterization based on laser-induced fluorescence [75]

fluorescence properties. Other pollutants and algal bloom patches can also be studied [76–78]. Using the blue-green transmission window of water, bathymetric measurements of sea depths can also be performed [79]. The field of laser-based hydrospheric monitoring is covered in [57–59].

Fluorescence monitoring techniques can obviously also be used for industrial purposes. Since oils possess very strong fluorescence properties the presence or absence of thin surface layers on e.g. sheet metal can be determined. Thus, surface cleanliness, corrosion protective measures etc. can be assessed with an industrial fluorosensor [80]. An illustration of fluorescence monitoring of the anti-rust oil application during coiling of sheet metal in a steel mill is given in Fig. 9 [80].

3. Medical Diagnostics

Lasers have found many applications in medicine. Most of them are focused around the heat generation in tissue caused by laser beams. Many surgical procedures can be performed with a focused laser beam at great advantage compared with a conventional scalpel [81]. The wavelength of the radiation will determine the penetration depth in tissue, with the CO_2 laser having the least, the argon-ion intermediate, and the Nd:YAG laser the greatest penetration. Various aspects of laser interaction with tissue are treated in [82, 83]. While the optical properties of tissue are important in laser surgery, spectroscopic aspects are much more dominant in another field: that of tumour localization and treatment using tumour-seeking and photosensitizing drugs. These methods are described in [84–86]. The principle of this technique, which is now being tested worldwide, is illustrated in Fig. 10. A sensitizer, most frequently hematoporphyrin derivative (HPD) is injected intravenously at low concentration. After a few days the agent is excreted from the body, but, for still only partially known reasons, a selective retention in tumours occurs. This tumour marking can now be utilized in two ways. The first application is tumour localization, utilizing the characteristic dual-peaked fluorescence distribution of HPD in the red spectral region. Fluorescence can be induced by a UV or violet laser beam (N_2 or krypton laser) and the fluorescence can be detected with an optical multichannel analyser using the same quartz fibre inserted into, e.g. the bronchi or the bladder. The fluorescence from the HPD-bearing tumour is seen superimposed on the tissue normal fluorescence (autofluorescence), which can also carry diagnostic information [87–89]. Fluorescence diagnostics of cancer tumours, including descriptions of imaging instrumentation are discussed in [90–94].

For tumour treatment a special type of laser-induced chemistry is employed. Red laser light at 630 nm, normally supplied from an argon-ion-pumped dye laser or a gold vapour laser is directed onto the tissue, which is reasonably transparent in this spectral region. The HPD molecules are excited and transfer their energy to ground-state oxygen molecules ($X^3\Sigma_g$ triplet oxygen) which are then excited to their $a^1\Delta_g$

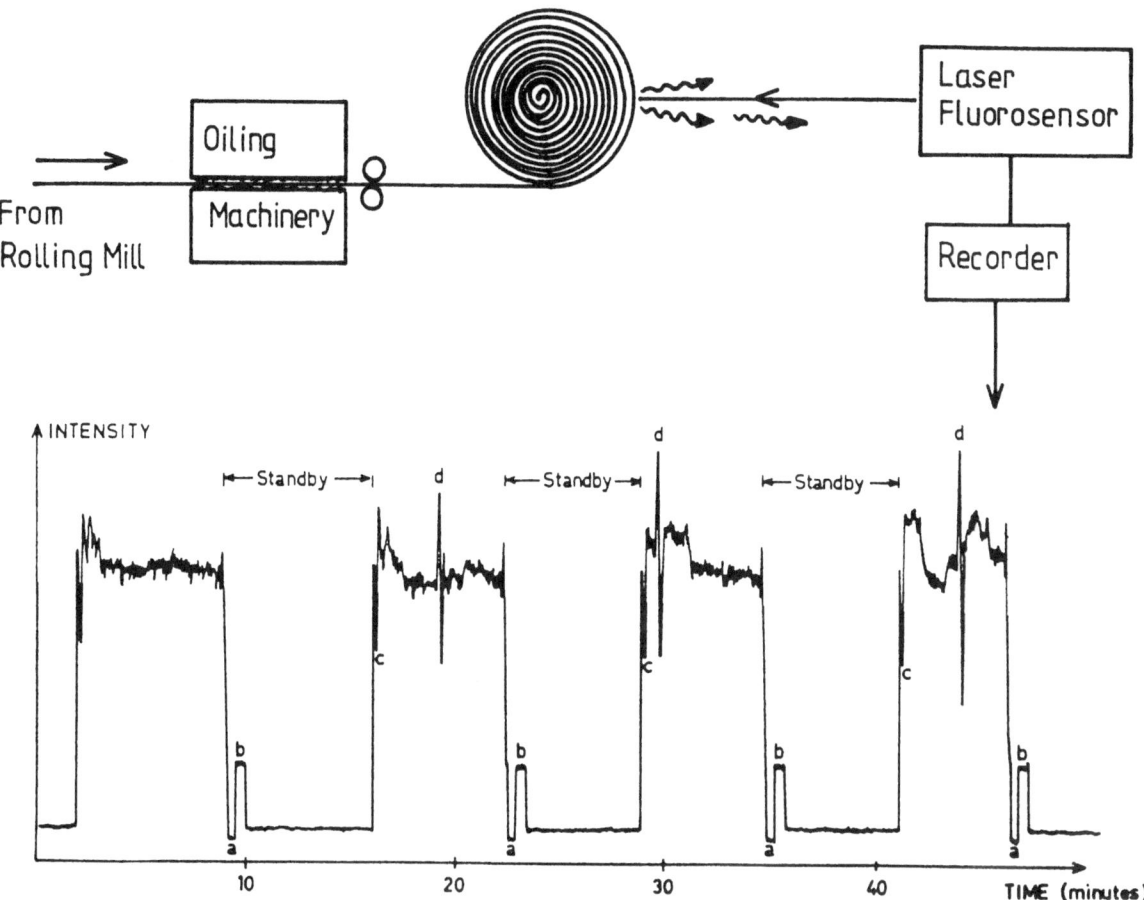

Fig. 9. Recording of the laser-induced fluorescence intensity from the rust-protective oil at the end of a cold-rolled strip steel line. The coiling of four steel rolls is monitored. At the points marked "d" the band was stopped to measure the width and thickness of the strip [80]

(singlet) state. Singlet oxygen is known to be strongly toxic to tissue, causing oxidation and necrosis. This process only occurs if HPD molecules are present, i.e. destruction of tumour cells is achieved. Superficial tumours can be irradiated from the surface, while thick tumours can be treated interstitially with fibres implanted in the tumour mass. Using fibre endoscopes, tumours can be detected and treated in the bronchi, bladder etc.

As an example of the fluorescence technique, data from a scan across a cancer tumour in a rat brain are shown in Fig. 11 [95]. The rat had been administered HPD 3 days before the investigation which was performed with a nitrogen laser (337 nm) and an optical multichannel analyser. The fluorescence spectrum from the tumour clearly shows the characteristic red HPD fluorescence peaks. It can also be noted, that the blue autofluorescence is much stronger in normal tissue than in tumour. Strong contrast enhancement is achieved by forming the ratio of the background free red HPD fluorescence and the blue autofluorescence.

Spectral data as a function of position along the scan across the tumour are given in the figure.

It is of great interest to develop equipment that can provide images of diseased tissue utilizing the techniques just described. A simple arrangement, providing imaging along a line, is illustrated in Fig. 12 [93]. Using cylinder lenses, a nitrogen laser beam is snaped into a 20 mm long, 1 mm wide line extending from the position of the tumour out into normal tissue. By adjusting the mirrors three images can be arranged side by side on an intensified linear array detector. A filter arrangement is used to filter the individual fluorescence images for passbands at 630 nm, 600 nm, and 470 nm. The result of such a procedure is shown in Fig. 13 for the case of a tumour in a rat, that had received HPD. The intensity in the 630 nm bandpass is higher in tumour than in muscle as expected. However, much better contrast is achieved by subtracting the non-specific red background light (intensity at 600 nm) and dividing by the blue light intensity, which is higher in normal tissue. In the lower part of

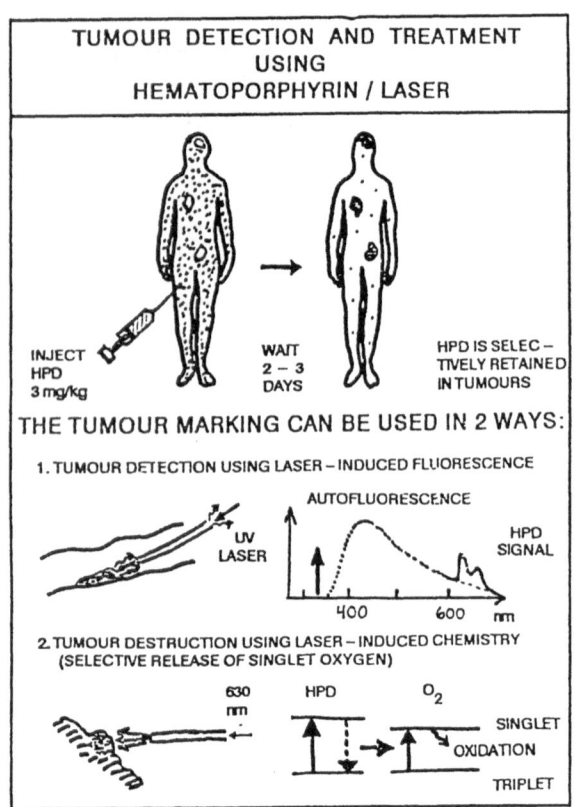

Fig. 10. Illustration of the principles for tumour detection and treatment using HpD combined with laser radiation [92, last ref.]

the figure the tumour demarcation function obtained from the three individual measurements is shown. Compared with the 630 nm curve the contrast between tumour and normal tissue has been enhanced by a factor of about 10. A system for two-dimensional imaging in 4 fluorescence colours is under construction at our laboratory [94].

The photodynamic action obtained by using a combination of HPD and laser light is illustrated in Fig. 14 [96], where a sequence of photographs is given showing the appearance of a basal cell cancer tumour before and at two times after irradiation. The patient had been administered 2 mg/kg bodyweight HPD 3 days before the irradiation with 60 J/cm². The tumour was fully eliminated by the procedure.

It should be noted, that the technique is still at the experimental stage and that no approval for its general use has been given in any country. Since there is also some accumulation of HPD in the skin, patients must be kept out of sunlight for at least four weeks, but this is the only complication known so far. A lot of work is being invested in the search for attractive new photosensitizers. Phtalocyanines and chlorines absorb at longer wavelengths than HPD does, allowing increased tissue penetration. The tumour demarcation abilities of different sensitizers using fluorescence detection are compared, e.g. in [97].

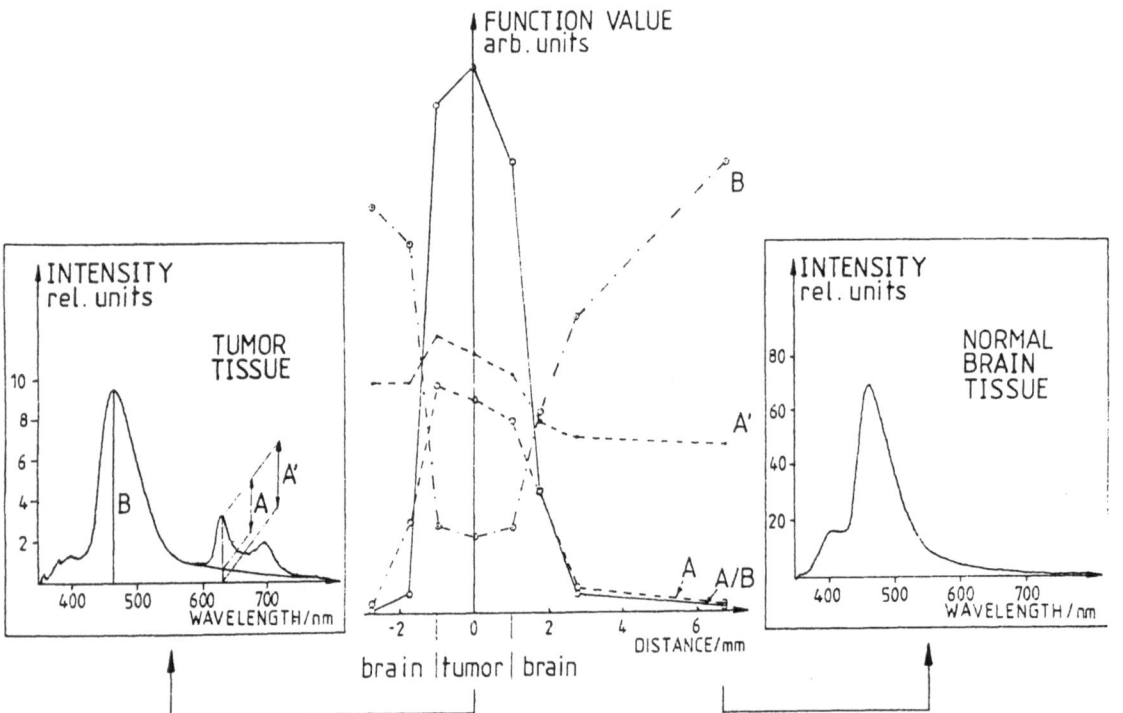

Fig. 11. Fluorescence data from a scan across a cancer tumour in the brain of a rat, that had been administered HPD 3 days earlier. Sample spectra for tumour and normal tissue are shown with evaluated parameters indicated. The tumour is clearly demarcated in the A and B functions and especially well in the A/B function [95]

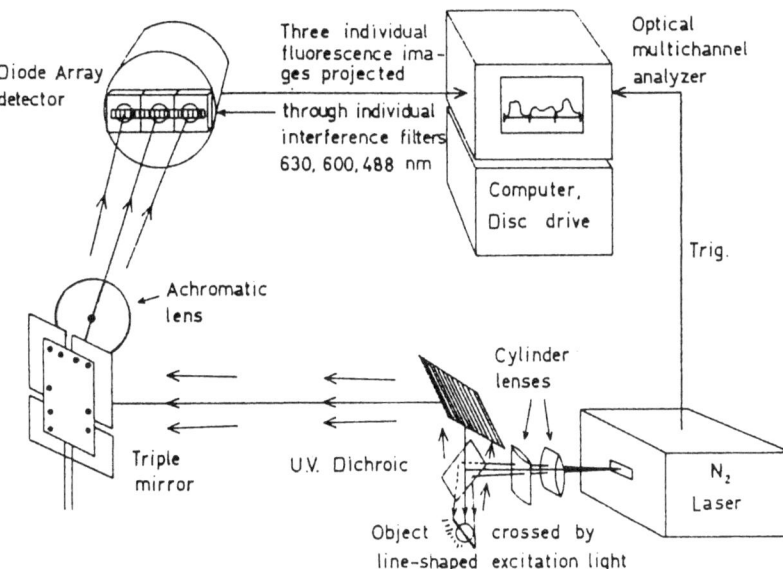

Fig. 12. Experimental arrangement for spatially resolved simultaneous recordings of laser-induced fluorescence in three spectral bands utilizing a single linear detector array [93]

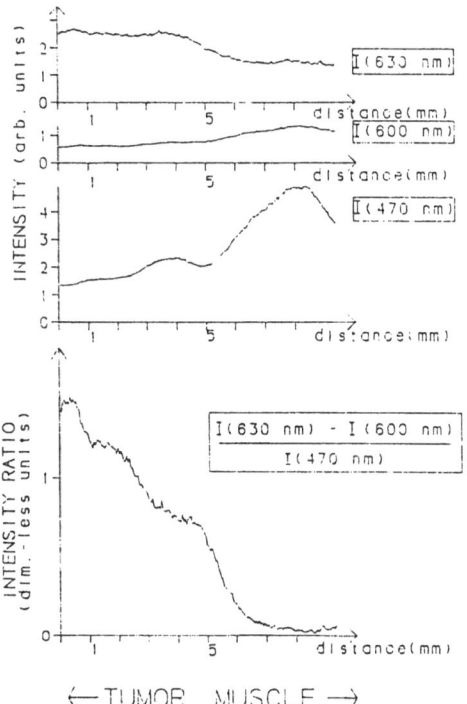

Fig. 13. Fluorescence imaging of a rat tumour and surrounding muscle in three colours using equipment of the type shown in Fig. 12. A contrast-enhanced image formed from the individual spectral images is given in the lower part of the figure [94]

Another emerging application of laser spectroscopy in medicine is the detection and characterization of atherosclerotic plaque. It would be of considerable value if vessels, e.g. the coronary arteries, could be cleared of obstructing plaque by laser evaporation using a fibre catheter. A great deal of research is being performed along these lines [98, 99]. One of the problems encountered is the risk of perforating the vessel with a high-power laser beam, inadvertently directed onto the normal wall instead of the plaque. Several groups are investigating the possibility of spectral identification of the tissue using laser-induced fluorescence (see, e.g. [100–103]). After spectral approval of the target the high-power pulse would then be fired through the same fibre. It seems that this technique has interesting possibilities. In Fig. 15 autofluorescence spectra of plaque and normal aortic wall are shown, exhibiting several marked differences [103].

The prospect of using fibre-optic procedures instead of major surgery in the fields of cancer and heart/circulation diseases is very attractive. It seems that optical spectroscopy based on lasers can play an important part in this process.

4. Conclusions

Applied laser spectroscopy is rapidly expanding the frontiers of advanced analysis and diagnostics. It has brought about many new possibilities in chemical analysis, diagnostics of reactive media, environmental and industrial monitoring as well as medical diagnostics. The main obstacles for the widespread application of the powerful techniques of laser spectroscopy are probably complexity of operation and maintenance of lasers, and to some extent high cost. Promising trends in this context are the rapid developments in semiconductor and solid state laser technology, new nonlinear optical materials and CCD (charge coupled device) detector technology.

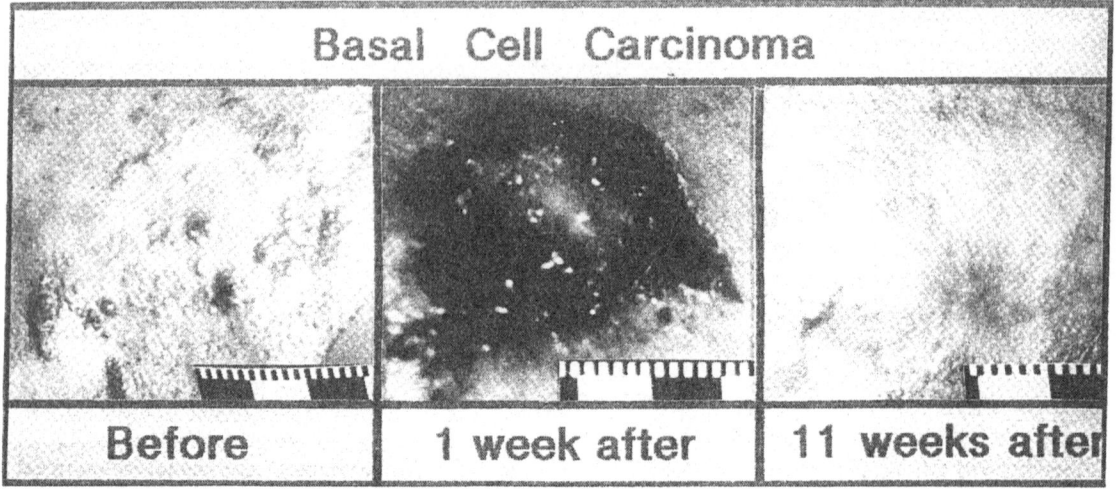

Fig. 14. Photographs of a basal-cell cancer tumour before, 1 week and 11 weeks after photodynamic therapy [96]

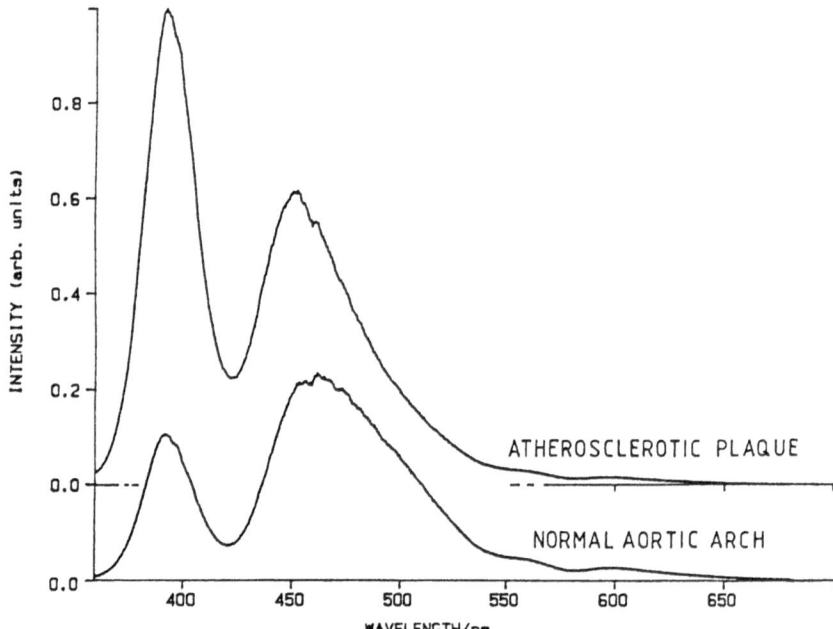

Fig. 15. Laser-induced fluorescence spectra for human normal aorta and atherosclerotic plaque [94]

Acknowledgements. A stimulating collaboration with a large number of colleagues and students inside and outside the Department of Physics is gratefully acknowledged. This work was supported by the Swedish Board for Technical Development (STU), the Swedish Board for Space Activities (DFR), the Swedish Environmental Protection Board (SNV), the Swedish Cancer Foundation (RMC) and the Wallenberg Foundation.

References

1. T.W. Hänsch, Y.R. Shen (eds.): *Laser Spectroscopy VII*, Springer Ser. Opt. Sci. Vol. **49** (Springer, Berlin, Heidelberg 1985)

2. W. Persson, S. Svanberg (eds.): *Laser Spectroscopy VIII*, Springer Ser. Opt. Sci. Vol. **55** (Springer, Berlin, Heidelberg 1987)

3. R.S. Van Dyck, Jr., E.N. Fortson (eds.): *Atomic Physics* **9** (World Scientific, Singapore 1985)

4. H. Narumi, I. Shimura (eds.): *Atomic Physics* **10** (North-Holland, Amsterdam 1987)

5. N. Omenetto (ed.): *Analytical Laser Spectroscopy* (Wiley, New York 1979)
   G.M. Hieftje, J.C. Travis, F.E. Lytle (eds.): *Lasers in Chemical Analysis* (Humana Press, Clifton, NJ 1981)
   V.S. Letokhov (ed.): *Laser Analytical Spectrochemistry* (Adam Hilger, Bristol 1985)
   E.H. Piepmeier (ed.): *Analytical Applications of Lasers* (Wiley Interscience, New York (1986)

6. K. Kleinermanns, J. Wolfrum: Angew. Chem. Int. Ed. Engl. **26**, 38 (1987)

7. V.S. Letokhov: *Non-Linear Laser Chemistry*, Springer Ser. Chem. Phys., Vol. **22** (Springer, Berlin, Heidelberg 1983); *Applications of Lasers to Industrial Chemistry*, SPIE Vol. **458** (SPIE, Bellham, Wash. D.C. 1984)

8. R.M. Osgood, S.R.J. Brueck, H.R. Schlossberg (eds.): *Laser Diagnostics and Photochemical Processing for Semiconductor Devices* (North-Holland, Amsterdam 1983)

9. D. Bäuerle (ed.): *Laser Processing and Diagnostics*, Springer Ser. Chem. Phys., Vol. 39 (Springer, Berlin, Heidelberg 1984)

10. D. Bäuerle: *Chemical Processing with Lasers*, Springer Ser. Mat. Sci., Vol. 1 (Springer, Berlin, Heidelberg 1986)

11. S. Svanberg: *Atomic and Molecular Spectroscopy*, Springer Ser. Opt. Sci., Vol. 56 (Springer, Berlin, Heidelberg, 1988) (to appear)

12. L.J. Radziemski, R.W. Solarz, J.A. Paisner (eds.): *Laser Spectroscopy and its Applications* (Dekker, New York 1987)

13. W.C. Gardiner, Jr.: Sci. Am. **246**, 110 (1982)

14. W.C. Gardiner, Jr. (ed.): *Combustion Chemistry* (Springer, Berlin, Heidelberg 1984)

15. D.R. Crosley (ed.): *Laser Probes for Combustion Chemistry*, ACS Symposium Series, Vol. 134 (Am. Chem. Soc., Washington 1980)

16. A.C. Eckbreth, P.A. Bonczyk, J.F. Verdiek: Prog. Energy Combust. Sci. **5**, 253 (1979)

17. J.H. Bechtel, A.R. Chraplyvy: Proc. IEEE **70**, 658 (1982)

18. J.H. Bechtel, C.J. Dasch, R.E. Teets: "*Combustion Research with Lasers*", *Laser Applications*, ed. by R.K. Erf, J.F. Ready (Academic, New York 1984)

19. T.D. McCay, J.A. Roux (eds.): Combustion Diagnostics by Nonintrusive Methods, Progress in Astronautics and Aeronautics Vol. **92**, 1983

20. K. Knapp, R.K. Hanson: Appl. Opt. **22**, 1980 (1983)
K.E. Bennett, G.W. Faris, R.L. Byer: Appl. Opt. **23**, 2678 (1984)

21. Special Issue on Computerized Tomography, Proc. IEEE **71**, 291 (1983); Special Issue on Industrial Applications of Computed Tomography and NMR Imaging, Appl. Opt. **24**, 23 (1985)

22. D.R. Crosley, G.P. Smith: Opt. Eng. **22**, 545 (1983)

23. R. Lucht: In [12], pp. 623

24. G. Kychakoff, R.D. Howe, R.K. Hanson: Appl. Opt. **23**, 704 (1984)
G. Kychakoff, K. Knapp, R.D. Howe, R.K. Hanson: AIAA J. **22**, 153 (1984)
R.K. Hanson: In Proc. Twenty-First Symposium on Combustion, Munich 1986 (The Combustion Institute, Pittsburg 1986)
B. Hiller, R.K. Hanson: Appl. Opt. **27**, 33 (1988)

25. M. Aldén, H. Edner, P. Grafström, H.M. Hertz, G. Holmstedt, T. Högberg, H. Lundberg, S. Svanberg, S. Wallin, W. Wendt, U. Westblom: In Lasers '86 (STS Press, McLean, VA 1985) p. 219

26. J.E.M. Goldsmith: In [2], pp. 337

27. A. D'Alessio, A. Di Lorenzo, A.F. Sarofim, F. Beretta, S. Masi, C. Venitozzi: The 15th (Int) Symposium on Combustion (The Combustion Institute, Pittsburg 1975) p. 1427
A. D'Alessio, A. Di Lorenzo, A. Borghese, F. Beretta, S. Masi: The 16th (Int) Symposium on Combustion (The Combustion Institute, Pittsburg 1977) p. 695

28. M. Lapp, C.M. Penney: In R.J.H. Clark, R.E. Hester (eds.): *Advances in Infrared and Raman Spectroscopy* (Heyden & Sons, London, 1977)

29. R.W. Dibble, A.R. Masri, R.W. Bilger: Combust. Flame **67**, 189 (1987)

30. A.C. Eckbreth, P.W. Schreiber: In *Chemical Applications of Non-Linear Raman Spectroscopy*, ed. by A.B. Harvey (Academic, New York 1981)
R.J. Hall, A.C. Eckbreth: In *Laser Applications*, Vol. 5, ed. by J.F. Ready, R.K. Erf (Academic, New York 1984)
A.C. Eckbreth: In [2], pp. 320

31. J.J. Valentini: In [12], pp. 507

32. M. Aldén, H. Edner, S. Svanberg: Phys. Scr. **27**, 29 (1983)

33. J. Pender, L. Hesselink: Opt. Lett. **10**, 264 (1985)

34. P. Ewart, S.V. O'Leary: Opt. Lett. **11**, 279 (1986)

35. A. Rose, G.J. Salamo, R. Gupta: Appl. Opt. **23**, 781 (1984)

36. H. Sonntag, A.C. Tam: Opt. Lett. **10**, 436 (1985)

37. G.W. Faris, R.L. Byer: Opt. Lett. **12**, 72 (1987); **12**, 155 (1987)

38. K. Tennel, G.J. Salomo, R. Gupta: Appl. Opt. **21**, 2133 (1982)

39. A.C. Tam: Rev. Mod. Phys. **58**, 381 (1986)

40. P. Camus (ed.): Optogalvanic Spectroscopy and its Applications, J. Physique Coll., Suppl. **44**, C7 (1983)

41. J.E.M. Goldsmith, J.E. Lawler: Contemp. Phys. **22**, 235 (1981)

42. J.E.M. Goldsmith: In [2]

43. M. Aldén, H.M. Hertz, S. Svanberg, S. Wallin: Appl. Opt. **23**, 3255 (1984)

44. C. Chan, J.W. Daily: Appl. Opt. **19**, 1963 (1980)
R.P. Lucht, N.N. Laurendeau, D.W. Sweeney: Appl. Opt. **21**, 3729 (1982)

45. M. Aldén, P. Grafström, H. Lundberg, S. Svanberg: Opt. Lett. **8**, 241 (1983)

46. D.A. Stephenson, R.J. Cattolica: Proc. 9th Int. Colloq. on Dynamics of Explosions and Reactive Systems, Poitier 1983

47. U. Westblom, S. Svanberg: Phys. Scr. **31**, 402 (1985)

48. G.C. Alessandretti, P. Violino: J. Phys. D **16**, 1583 (1983)

49. M. Aldén, S. Wallin: Appl. Opt. **24**, 3434 (1985)

50. A.C. Eckbreth, T.J. Andersson: Appl. Opt. **24**, 2731 (1985); Opt. Lett. **11**, 496 (1986)
M. Aldén, P.E. Bengtsson, H. Edner: Appl. Opt. **25**, 4493 (1986)

51. D.A. Greenhalgh, S.T. Whittley: Appl. Opt. **24**, 907 (1985)
R.J. Hall, D.A. Greenhalgh: J. Opt. Soc. Am. B **3**, 1637 (1986)
S. Kröll, M. Aldén, T. Berglind, R.J. Hall: Appl. Opt. **26**, 1068 (1987)
D.R. Snelling, T. Parameswaran, G.J. Smallwood: Appl. Opt. **26**, 4298 (1987)

52. W. Bach, J. Pankrath, W. Kellogg (eds.): *Man's Impact on Climate* (Elsevier, Amsterdam 1979)

53. R. Revelle: Sci. Am. **247**, 35 (1982)

54. J.H. Seinfeld: *Atmospheric Chemistry and Physics of Air Pollution* (Wiley, New York 1986)

55. R.P. Wayne: *Chemistry of Atmospheres* (Clarendon Press, Oxford 1985)

56. D.A. Killinger, A. Mooradian (eds.): *Optical and Laser Remote Sensing*, Springer Ser. Opt. Sci., Vol. 39 (Springer, Berlin, Heidelberg 1983)

57. R.M. Measures: *Laser Remote Sensing: Fundamentals and Applications* (Wiley, New York 1984)
R.M. Measures: In [5a]

58. E.D. Hinkley (ed.): *Laser Monitoring of the Atmosphere*, Topics Appl. Phys., Vol. 14 (Springer, Berlin, Heidelberg 1976)

59. S. Svanberg: "Lasers as Probes for Air and Sea", Contemp. Phys. **21**, 541 (1980)
S. Svanberg: Fundamentals of Atmospheric Spectroscopy. *Surveillance of Environmental Pollution and Resources by Electromagnetic Waves*, ed. by T. Lund (Reidel, Dordrecht 1978)

60. W.B. Grant: *Laser Remote Sensing Techniques*, in [12] pp. 565

61. U. Platt, D. Perner, H.W. Pätz: J. Geophys. Res. **84**, 6329 (1979)
    U. Platt, D. Perner: In [56]
62. H. Edner, A. Sunesson, S. Svanberg, L. Unéus, S. Wallin: Appl. Opt. **25**, 403 (1986)
63. M.L. Chanin: In [56]
    C. Granier, G. Megie: Planet. Space Sci. **30**, 169 (1982)
    C. Granier, J.P. Jegou, G. Megie: Geophys. Res. Lett. **12**, 655 (1985)
    K.H. Fricke, U. v. Zahn: J. Atm. Terr. Phys. **47**, 499 (1985); Geophys. Res. Lett. **14**, 76 (1987)
64. M.P. McCormick, T.J. Swisser, W.H. Fuller, W.H. Hunt, M.T. Osborn: Geofisica Internacional **23-2**, 187 (1984)
65. H. Edner, K. Fredriksson, A. Sunesson, S. Svanberg, L. Unéus, W. Wendt: Appl. Opt. **26**, 4330 (1987)
66. K. Fredriksson, B. Galle, K. Nyström, S. Svanberg: Appl. Opt. **20**, 4181 (1981)
    K. Fredriksson, S. Svanberg: In [56]
    K. Fredriksson, H.M. Hertz: Appl. Opt. **23**, 1403 (1984)
    A.L. Egebäck, K. Fredriksson, H.M. Hertz: Appl. Opt. **23**, 722 (1984)
67. H. Edner, G.W. Faris, A. Sunesson, S. Svanberg: Appl. Opt. (in press)
68. Q. Bristow, I.R. Jonason: Can. Min. J. **93**, 39 (1972)
69. A.F. Jepsen: *Measurements of Mercury Vapor in the Atmosphere*. In Advances in Chemistry Series, Vol. **123** (Am. Chem. Soc., Washington D.C. 1973) pp. 80
70. J.C. Varekamp, P.R. Buseck: Nature **293**, 555 (1981)
71. G. Megie, G. Ancellet, J. Pelon: Appl. Opt. **24**, 3454 (1985)
72. O. Uchino, M. Togunaga, M. Maeda, Y. Miyazoe: Opt. Lett. **8**, 347 (1983)
73. J. Werner, K.W. Rothe, H. Walther: Appl. Phys. B **32**, 113 (1983)
74. H. Edner, S. Svanberg, L. Unéus, W. Wendt: Opt. Lett. **9**, 493 (1984)
75. P.S. Andersson, S. Montán, S. Svanberg: Appl. Phys. B **44**, 19 (1987)
76. F.E. Hoge, R.N. Swift: Appl. Opt. **20**, 3197 (1981); **25**, 48 (1986)
77. R.A. O'Neill, L. Buja-Bijunas, D.M. Rayner: Appl. Opt. **19**, 863 (1980)
78. G.A. Capelle, L.A. Franks, D.A. Jessup: Appl. Opt. **22**, 3382 (1983)
79. H.H. Kim: Appl. Opt. **16**, 45 (1977)
    F.E. Hoge, R.N. Swift, E.B. Frederick: Appl. Opt. **19**, 871 (1980)
80. S. Montán, S. Svanberg: Appl. Phys. B **38**, 241 (1985); L.I.A. ICALEO **47**, 153 (1985)
81. L. Goldman (ed.): *The Biomedical Laser: Technology and Clinical Applications* (Springer, Berlin, Heidelberg 1981)
82. J.L. Boulnois: Lasers Med. Sci. **1**, 47 (1986)
83. J.A. Parrish, T.F. Deutsch: IEEE J. QE-**20**, 1386 (1984)
84. T.J. Dougherty: In *CRC Critical Reviews in Oncology/Hematology*, ed. by S. Davis (CRC, Boca Raton 1984)
85. Y. Hayata, T.J. Dougherty (eds.): *Lasers and Hematoporphyrin Derivative in Cancer* (Ikaku-Shoin, Tokyo 1983)
86. Y. Hayata (ed.): Proceedings of the First International Conference on the Application of Photosensitization for Diagnosis and Treatment. Tokyo 1986 (in press)
87. Y. Yanming, Y. Yuanlong, L. Yufen, L. Fuming: CLEO'85 Technical Digest, p. 84 and previous references in Chinese
88. R.R. Alfano, B.T. Darayash, J. Cordero, P. Tomashefsky, F.W. Longo, M.A. Alfano: IEEE J. QE-**20**, 1507 (1984)
    D.B. Tata, M. Foresti, J. Cordero, P. Thomachevsky, M.A. Alfano, R.R. Alfano: Biophys. J. **50**, 463 (1986)
    R.R. Alfano, W. Lam, H. Zarabbi, M.A. Alfano, J. Cordero, D. Tata, C. Swenberg: IEEE J. QE-**20**, 1512 (1984)
89. P.S. Andersson, E. Kjellén, S. Montán, K. Svanberg, S. Svanberg: Lasers Med. Sci. **2**, 41 (1986)
    S. Montán, L.G. Strömblad: Lasers Life Sci. **1**, 275 (1987)
90. A.E. Profio, D.R. Doiron, O.J. Balchum, G.C. Huth: Med. Phys. **10**, 35 (1983)
91. H. Kato, D.A. Cortese: Clin. Chest Med. **6**, 237 (1985)
92. J. Ankerst, S. Montán, K. Svanberg, S. Svanberg: Appl. Spectr. **38**, 890 (1984)
    K. Svanberg, E. Kjellén, J. Ankerst, S. Montán, E. Sjöholm, S. Svanberg: Cancer Res. **46**, 3803 (1986)
    P.S. Andersson, S.E. Karlsson, S. Montán, T. Persson, S. Svanberg, S. Tapper: Med. Phys. **14**, 633 (1987)
    S. Svanberg: Phys. Scr. T **19**, 469 (1987)
93. S. Montán, K. Svanberg, S. Svanberg: Opt. Lett. **10**, 56 (1985)
94. P.S. Andersson, S. Montán, S. Svanberg: IEEE J. QE-**23**, 1798 (1987)
95. S. Andersson-Engels, A. Brun, E. Kjéllen, L.G. Salford, L.G. Strömblad, K. Svanberg, S. Svanberg: J. Neurosurg. (to appear)
96. S. Andersson-Engels, J. Johansson, E. Kjellén, D. Killander, L.O. Svaasand, K. Svanberg, S. Svanberg: L.I.A. ICALEO **60**, 67 (1987)
97. S. Andersson-Engels, J. Ankerst, S. Montán, K. Svanberg, S. Svanberg: Lasers in Med. Sci. (in press)
    S. Andersson-Engels, J. Ankerst, J. Johansson, K. Svanberg, S. Svanberg: SPIE Conf. Proc. 1988 (in press)
98. J.M. Isner, R.H. Clarke: IEEE J. QE-**20**, 1406 (1984)
    J.M. Isner, P.G. Steg, R.H. Clarke: IEEE J. QE-**23**, 1756 (1987)
99. M.R. Prince, T.F. Deutsch, M.M. Mathews-Roth, R. Margolis, J.A. Parrish, A.R. Oseroff: J. Clin. Invest. **78**, 295 (1986)
100. C. Kittrell, R.L. Willett, C. de los Santos-Pacheo, N.B. Ratliff, J.R. Kramer, E.G. Malk, M.S. Feld: Appl. Opt. **24**, 2280 (1985)
     R.N. Cothren, G.B. Hayes, J.R. Kramer, B. Sacks, C. Kittrell, M.S. Feld: Lasers Life Sci. **1**, 1 (1986)
101. A.A. Orayevski, V.S. Letokhov, S.E. Ragimov, V.G. Omelyanenko, A.A. Belyaev, B.V. Shokonin, R.S. Akchurin: Lasers Life Sci. (in press)
102. M. Sartori, R. Sauerbrey, S. Kubodera, F.K. Tittel, R. Roberts, P.D. Henry: IEEE J. QE-**23**, 1794 (1987)
103. P.S. Andersson, A. Gustafson, U. Stenram, K. Svanberg, S. Svanberg: Lasers Med. Sci. **2**, 261 (1987)

Responsible for Advertisements: E. Lückermann, G. Probst, Heidelberger Platz 3, D-1000 Berlin 33, Tel. (030) 8207-0, Telex 01-85411
Responsible for the Text: H.K.V. Lotsch, Tiergartenstrasse 17, D-6900 Heidelberg. Printers: Brühlsche Universitätsdruckerei, Giessen